Marx Joyce
Abbott Hardy Cooper Austen
Defoe Melville Montaigne Chesterton Emerson Hugo
Machiavelli Eliot
Stoker Christie Haggard Molière Grimm
Wilde Carroll Maupassant Byron
Garnett Engels Schiller
Goethe Fitzgerald Hawthorne Smith Kafka
Cotton Einstein Dostoyevsky Hall
Baum Kipling Doyle Willis
Leslie Henry Nietzsche
Dumas Flaubert Turgenev Balzac
Stockton Vatsyayana Crane
Burroughs Verne
Curtis Tocqueville Whitman Vinci
Homer Widger Tolstoy Gogol Busch
Darwin Thoreau
Potter Freud Zola Twain Scott
Kant Jowett Lawrence Plato Harte
Stevenson Dickens Hesse
Andersen Burton
London Descartes Cervantes Cooke
Poe Aristotle Voltaire
Hale James Hastings
Bunner Shakespeare Irving
Richter Chambers
Doré da Benedict
Dante Shaw Pushkin Alcott
Swift Chekhov Wodehouse
Newton

tredition®

tredition was established in 2006 by Sandra Latusseck and Soenke Schulz. Based in Hamburg, Germany, tredition offers publishing solutions to authors and publishing houses, combined with worldwide distribution of printed and digital book content. tredition is uniquely positioned to enable authors and publishing houses to create books on their own terms and without conventional manufacturing risks.

For more information please visit: www.tredition.com

TREDITION CLASSICS

This book is part of the TREDITION CLASSICS series. The creators of this series are united by passion for literature and driven by the intention of making all public domain books available in printed format again - worldwide. Most TREDITION CLASSICS titles have been out of print and off the bookstore shelves for decades. At tredition we believe that a great book never goes out of style and that its value is eternal. Several mostly non-profit literature projects provide content to tredition. To support their good work, tredition donates a portion of the proceeds from each sold copy. As a reader of a TREDITION CLASSICS book, you support our mission to save many of the amazing works of world literature from oblivion. See all available books at www.tredition.com.

Project Gutenberg

The content for this book has been graciously provided by Project Gutenberg. Project Gutenberg is a non-profit organization founded by Michael Hart in 1971 at the University of Illinois. The mission of Project Gutenberg is simple: To encourage the creation and distribution of eBooks. Project Gutenberg is the first and largest collection of public domain eBooks.

A New Orchard And Garden or, The best way for planting, grafting, and to make any ground good, for a rich Orchard: Particularly in the North and generally for the whole kingdome of England

William Lawson

Imprint

This book is part of TREDITION CLASSICS

Author: William Lawson
Cover design: Buchgut, Berlin – Germany

Publisher: tredition GmbH, Hamburg - Germany
ISBN: 978-3-8472-1621-6

www.tredition.com
www.tredition.de

Copyright:
The content of this book is sourced from the public domain.

The intention of the TREDITION CLASSICS series is to make world literature in the public domain available in printed format. Literary enthusiasts and organizations, such as Project Gutenberg, worldwide have scanned and digitally edited the original texts. tredition has subsequently formatted and redesigned the content into a modern reading layout. Therefore, we cannot guarantee the exact reproduction of the original format of a particular historic edition. Please also note that no modifications have been made to the spelling, therefore it may differ from the orthography used today.

A
NEVV ORCHARD
AND GARDEN

OR

The best way for planting, grafting, and to make
any ground good, for a rich Orchard: Particularly in the North,
and generally for the whole kingdome of *England*, as in nature,
reason, situation, and all probabilitie, may and doth appeare.
With the Country Housewifes Garden for hearbes of common vse:
their vertues, seasons, profits, ornaments, varietie of knots, models
for trees, and plots for the best ordering of Grounds and Walkes.
AS ALSO,
The Husbandry of Bees, with their seuerall vses and annoyances
being the experience of 48 yeares labour, and now the second time
corrected
and much enlarged, by *William Lawson*.

Whereunto is newly added the Art of propagating Plants, with the
true
ordering of all manner of Fruits, in their gathering, carrying home, &
preseruation.

Skill and paines bring fruitfull gaines.

Nemo sibi natus.

LONDON,
Printed by *Nicholas Okes* for Iohn Harison, at the golden
Vnicorne in Pater-noster-row. 1631.

TO THE RIGHT WORSHIPFVLL Sir Henry Belosses, Knight and Baronet,

Worthy Sir,

When in many yeeres by long experience I had furnished this my Northerne Orchard and Countrey Garden with needfull plants and vsefull hearbes, I did impart the view thereof to my friends, who resorted to me to conferre in matters of that nature, they did see it, and seeing it desired, and I must not denie now the publishing of it (which then I allotted to my priuate delight) for the publike profit of others. Wherefore, though I could pleade custome the ordinarie excuse of all Writers, to chuse a Patron and Protector of their Workes, and so shroud my selfe from scandall vnder your honourable fauour, yet haue I certaine reasons to excuse this my presumption: First, the many courtesies you haue vouchsafed me. Secondly, your delightfull skill in matters of this nature. Thirdly, the profit which I receiued from your learned discourse of Fruit-trees.

Fourthly, your animating and assisting of others to such endeuours. Last of all, the rare worke of your owne in this kind: all which to publish vnder your protection, I haue aduentured (as you see). Vouchsafe it therefore entertainement, I pray you, and I hope you shall finde it not the vnprofitablest seruant of your retinue: for when your serious employments are ouerpassed, it may interpose some commoditie, and raise your contentment out of varietie.

Your Worships
most bounden,

William Lavvson.

THE PREFACE to all well minded.

Art hath her first originall out of experience, which therefore is called the Schoole-mistresse of fooles, because she teacheth infallibly, and plainely, as drawing her knowledge out of the course of Nature, (which neuer failes in the generall) by the senses, feelingly apprehending, and comparing (with the helpe of the minde) the workes of nature; and as in all other things naturall, so especially in Trees; for what is Art more then a prouident and skilfull Collectrix of the faults of Nature in particular workes, apprehended by the senses? As when good ground naturally brings forth thistles, trees stand too thicke, or too thin, or disorderly, or (without dressing) put forth vnprofitable suckers, and suchlike. All which and a thousand more, Art reformeth, being taught by experience: and therefore must we count that Art the surest, that stands vpon experimentall rules, gathered by the rule of reason (not conceit) of all other rules the surest.

Whereupon haue I of my meere and sole experience, without respect to any former written Treatise, gathered these rules, and set them downe in writing, not daring to hide the least talent giuen me of my Lord and Master in Heauen: neither is this iniurious to any, though it differ from the common opinion in diuers points, to make it knowne to others, what good I haue found out in this facultie by long triall and experience. I confesse freely my want of curious skill in the Art of planting. And I admire and praise *Plinie, Aristotle, Virgil, Cicero,* and many others for wit and iudgement in this kind, and leaue them to their times, manner, and seuerall Countries.

I am not determined (neither can I worthily) to set forth the praises of this Art: how some, and not a few, euen of the best, haue accounted it a chiefe part of earthly happinesse, to haue faire and pleasant Orchards, as in *Hesperia* and *Thessaly,* how all with one consent agree, that it is a chiefe part of Husbandry (as *Tully de senectute*) and Husbandry maintaines the world; how ancient, how profitable, how pleasant it is, how many secrets of nature it doth containe, how loued, how much practised in the best places, and of the best: This hath already beene done by many. I only aime at the common good. *I* delight not in curious conceits, as planting and graffing with the root vpwards, inoculating Roses on Thornes, and such like, although I haue heard of diuers prooued some, and read of moe.

The Stationer hath (as being most desirous with me, to further the common good) bestowed much cost and care in hauing the Knots and Models by the best Artizan cut in great varietie, that nothing might be any way wanting to satisfie the curious desire of those that would make vse of this Booke.

And I shew a plaine and sure way of planting, which I haue found good by 48. yeeres (and moe) experience in the North part of *England*: I preiudicate and enuie none, wishing yet all to abstaine from maligning that good (to them vnknowne) which is well intended. Farewell.

Thine, for thy good, *W. L.*

THE BEST, SVRE
AND READIEST VVAY
to make a good *Orchard* and *Garden*.

Chapter. 1.
Of the Gardner, and his Wages.

Religious. Whosoeuer desireth & endeauoureth to haue a pleasant, and profitable Orchard, must (if he be able) prouide himselfe of a Fruicterer, religious, honest, skilful in that faculty, & therwithall painfull: By religious, I meane (because many think religion but a fashion or custome to go to Church) maintaining, & cherishing things religious: as Schooles of learning, Churches, Tythes, Churchgoods, & rights; and aboue all things, Gods word, & the Preachers thereof, so much as he is able, practising prayers, comfortable conference, mutuall instruction to edifie, almes, and other works of Charity, and all out of a good conscience.

Honest. Honesty in a Gardner, will grace your Garden, and all your house, and helpe to stay vnbridled Seruingmen, giuing offence to none, not calling your name into question by dishonest acts, nor infecting your family by euill counsell or example. For there is no plague so infectious as Popery and knauery, he will not purloine your profit, nor hinder your pleasures.

Skilfull. Concerning his skill, he must not be a Scolist, to make shew or take in hand that, which he cannot performe, especially in so weighty a thing as an Orchard: than the which, there can be no humane thing more excellent, either for pleasure or profit, as shall (God willing) be proued in the treatise following. And what an hinderance shall it be, not onely to the owner, but to the common good, that the vnspeakeble benefit of many hundred yeeres shall be lost, by the audacious attempt of an vnskilfull Arborist.

Painfull. The Gardner had not need be an idle, or lazie Lubber, for to your Orchard being a matter of such moment, will not prosper. There will euer be some thing to doe. Weedes are alwaies growing. The great mother of all liuing Creatures, the Earth, is full of seed in her bowels, and any stirring giues them heat of Sunne, and being laid neere day, they grow: Mowles worke daily, though not alwaies alike. Winter herbes at all times will grow (except in extreame frost.)

In Winter your young trees and herbes would be lightned of snow, and your Allyes cleansed: drifts of snow will set Deere, Hares, and Conyes, and other noysome beasts ouer your walles & hedges, into your Orchard. When Summer cloathes your borders with greene and peckled colours, your Gardner must dresse his hedges, and antike workes: watch his Bees, and hiue them: distill his Roses and other herbes. Now begins Summer Fruit to ripe, and craue your hand to pull them. If he haue a Garden (as he must need) to keepe, you must needs allow him good helpe, to end his labours which are endlesse, for no one man is sufficient for these things.

Wages. Such a Gardner as will conscionably, quietly and patiently, trauell in your Orchard, God shall crowne the labours of his hands with ioyfulnesse, and make the clouds drop fatnesse vpon your trees, he will prouoke your loue, and earne his wages, and fees belonging to his place: The house being serued, fallen fruite, superfluity of herbes, and flowers, seedes, grasses, sets, and besides all other of that fruit which your bountifull hand shall reward him withall, will much augment his wages, and the profit of your bees will pay you backe againe.

If you be not able, nor willing to hire a gardner, keepe your profits to your selfe, but then you must take all the pains: And for that purpose (if you want this faculty) to instruct you, haue I vndertaken these labours, and gathered these rules, but chiefly respecting my Countries good.

Chap. 2.
Of the soyle.

Kinds of trees. Fruit-trees most common, and meetest for our Northerne Countries: (as Apples, Peares, Cheries, Filberds, red and white Plummes, Damsons, and Bulles,) for we meddle not with Apricockes nor Peaches, nor scarcely with Quinces, which will not like in our cold parts, vnlesse they be helped with some reflex of Sunne, or other like meanes, nor with bushes, bearing berries, as Barberies, Goose-berries, or Grosers, Raspe-berries, and such like, though the Barbery be wholesome, and the tree may be made great: doe require (as all other trees doe) a blacke, fat, mellow, cleane and well tempered soyle, wherein they may gather plenty of good sap.

Some thinke the Hasell would haue a chanily rocke, and the sallow, and eller a waterish marish. Soyle. The soile is made better by deluing, and other meanes, being well melted, and the wildnesse of the earth and weedes (for euery thing subiect to man, and seruing his vse (not well ordered) is by nature subiect to the curse,) is killed by frosts and drought, by fallowing and laying on heapes, and if it be wild earth, with burning.

Barren earth. If your ground be barren (for some are forced to make an Orchard of barren ground) make a pit three quarters deepe, and two yards wide, and round in such places, where you would set your trees, and fill the same with fat, pure, and mellow earth, one whole foot higher then your Soile, and therein set your Plant. For who is able to manure an whole Orchard plot, if it be barren? But if you determine to manure the whole site, this is your way: digge a trench halfe a yard deepe, all along the lower (if there be a lower) side of your Orchard plot, casting vp all the earth on the inner side, and fill the same with good short, hot, & tender muck, and make such another Trench, and fill the same as the first, and so the third, and so through out your ground. And by this meanes your plot shall be fertile for your life. But be sure you set your trees, neither in dung nor barren earth.

Plaine. Your ground must be plaine, that it may receiue, and keepe moysture, not onely the raine falling thereon, but also water cast vpon it, or descending from higher ground by sluices, Conduits, &c. Moyst. For I account moisture in Summer very needfull in the soile of trees, & drought in Winter. Prouided, that the ground neither be boggy, nor the inundation be past 24. houres at any time, and but twice in the whole Summer, and so oft in the Winter. Therefore if your plot be in a Banke, or haue a descent, make Trenches by degrees, Allyes, Walkes, and such like, so as the Water may be stayed from passage. And if too much water be any hinderance to your walks (for dry walkes doe well become an Orchard, and an Orchard them:) raise your walkes with earth first, and then with stones, as bigge as Walnuts: and lastly, with grauell. In Summer you need not doubt too much water from heauen, either to hurt the health of your body, or of your trees. And if ouerflowing molest you after one day, auoid it then by deepe trenching.

Some for this purpose dig the soile of their Orchard to receiue moisture, which I cannot approue: for the roots with digging are oftentimes hurt, and especially being digged by some vnskilfull seruant: For the Gardiner cannot doe all himselfe. And moreouer, the roots of Apples & Peares being laid neere day, with the heate of the Sun, will put forth suckers, which are a great hinderance, and sometimes with euill guiding, the destruction of trees, vnlesse the deluing be very shallow, and the ground laid very leuell againe. Cherries and Plummes without deluing, will hardly or neuer (after twenty yeares) be kept from such suckers, nor aspes.

Grasse. Grasse also is thought needfull for moisture, so you let it not touch the roots of your trees: for it will breed mosse, and the boall of your tree neere the earth would haue the comfort of the Sunne and Ayre.

Some take their ground to be too moist when it is not so, by reason of waters standing thereon, for except in soure marshes, springs, and continuall ouerflowings, no earth can be too moyst. Sandy & fat earth will auoid all water falling by receit. Indeed a stiffe clay will not receiue the water, and therefore if it be grassie or plaine, especially hollow, the water will abide, and it wil seeme waterish, when the fault is in the want of manuring, and other good dressing.

Naturally plaine. This plainnesse which we require, had need be naturall, because to force an vneuen ground will destroy the fatnesse. For euery soile hath his crust next day wherein trees and herbes put their roots, and whence they draw their sap, which is the best of the soile, and made fertile with heat and cold, moisture and drought, and vnder which by reason of the want of the said temperature, by the said foure qualities, no tree nor herbe (in a manner) will or can put root. As may be seene if in digging your ground, you take the weeds of most growth: as grasse or docks, (which will grow though they lie vpon the earth bare) yet bury them vnder the crust, and they will surely dye and perish, & become manure to your ground. This crust is not past 15. or 18. inches deepe in good ground, in other grounds lesse. Crust of the earth. Hereby appeares the fault of forced plaines, viz. your crust in the lower parts, is couered with the crust of the higher parts, and both with worse earth: your heights hauing the crust taken away, are become meerely barren: so that either you must force a new crust, or haue an euill soile. And be sure you leuell, before you plant, lest you be forced to remoue, or hurt your plants by digging, and casting amongst their roots. Your ground must be cleered as much as you may of stones, and grauell, walls, hedges, bushes, & other weeds.

Chap. 3.
Of the Site.

There is no difference, that I find betwixt the necessity of a good soile, and a good site of an Orchard. For a good soile (as is before described) cannot want a good site, and if it do, the fruit cannot be good, and a good site will much mend an euill soile. Low and neere a Riuer. The best site is in low grounds, (and if you can) neere vnto a Riuer. High grounds are not naturally fat.

And if they haue any fatnesse by mans hand, the very descent in time doth wash it away. It is with grounds in this case as it is with men in a common wealth. Much will haue more: and once poore, seldome or neuer rich. The raine will scind, and wash, and the wind will blow fatnesse from the heights to the hollowes, where it will abide, and fatten the earth though it were barren before.

Hence it is, that we haue seldome any plaine grounds, and low, barren: and as seldome any heights naturally fertill. It is vnspeakeable, what fatnesse is brought to low grounds by inundations of waters. Neither did I euer know any barren ground in a low plaine by a Riuer side. The goodnesse of the soile in *Howle* or *Hollowdernes*, in *York-shire*, is well knowne to all that know the Riuer *Humber*, and the huge bulkes of their Cattell there. By estimation of them that haue seene the low grounds in *Holland* and *Zealand* they farre surpasse the most Countries in *Europe* for fruitfulnesse, and only because they lie so low. Psal. 1. 3.
Ezek. 17. 8.
Eccl. 39. 17. The world cannot compare with *Ægypt*, for fertility, so farre as *Nilus* doth ouer flow his bankes. So that a fitter place cannot be chosen for an Orchard, then a low plaine by a riuer side. For besides the fatnesse which the water brings, if any cloudy mist or raine be stirring, it commonly falls downe to, and followes the

course of the Riuer. And where see we greater trees of bulke and bough, then standing on or neere the waters side? If you aske why the plaines in *Holderns*, and such countries are destitute of woods? I answer that men and cattell (that haue put trees thence, from out of Plaines to void corners) are better then trees. Neither are those places without trees. Mr. *Markham*. Our old fathers can tel vs, how woods are decaied, & people in the roomth of trees   multiplied. I haue stood somwhat long in this poynt, because some do condemne a moist soile for fruit-trees.

Winds.

Chap. 13. A low ground is good to auoide the danger of winds, both for shaking downe your vnripe fruite. Trees the most (that I know) being loaden with wood, for want of proyning, and growing high, by the vnskilfulnesse of the Arborist, must needes be in continuall danger of the South-west, West, and North west winds, especially in *September* and *March*, when the aire is most temperate from extreme heat, and cold, which are deadly enemies to great winds. Wherefore chuse your ground low: Or if you be forced to plant in a higher ground, let high and strong wals, houses, and trees, as wallnuts, plane trees, Okes, and Ashes, placed in good order, be your fence for winds.

The sucken of your dwelling house, descending into your orchard, if it be cleanly conueyed, is good.

Sunne. The Sunne, in some sort, is the life of the world. It maketh proud growth, and ripens kindly, and speedily, according to the golden tearme: *Annus fructificat, non tellus*. Therefore in the countries, neerer approching the Zodiake, the Sunnes habitation, they haue better, and sooner ripe fruite, then we that dwell in these frozen parts.

Trees against a wall. This prouoketh most of our great Arborists, to plant Apricockes, Cherries and Peaches, by a wall, and with tackes, and other meanes to spread them vpon, and fasten them to a wall, to haue the benefit of the immoderate reflexe of the Sunne, which is commendable, for the hauing of faire, good & soone ripe fruit. But let them know it is more hurtfull to their trees then the benefit they reape therby: as not suffering a tree to liue the tenth part of his age. It helpes Gardners to worke, for first the   wall hinders the roots, because into a dry and hard wall of earth or stone a tree will not, nor cannot put any root to profit, but especially it stops the passage of sap, whereby the barke is wounded, & the wood, & diseases grow, so that the tree becomes short of life. For as in the body of a man, the leaning or lying on some member, wherby the course of bloud is stopt, makes that member as it were dead for the time, till the bloud returne to his course, and I thinke, if that stopping should continue any time, the member would perish for want of bloud (for the life is in the bloud) and so endanger the body: so the sap is the life of the tree, as the bloud is to mans body: neither doth the tree in winter (as is supposed) want his sap, no more then mans body his bloud, which in winter, and time of sleep draws inward. So that the dead time of winter, to a tree, is but a night of rest: for the tree at all times, euen in winter is nourished with sap, & groweth as well as mans body. The chilling cold may well some little time stay, or hinder the proud course of the sap, but so little & so short a time, that in calme & mild season, euen in the depth of winter, if you marke it, you may easily perceiue, the sap to put out, and your trees to increase their buds, which were formed in the summer before, & may easily be discerned: for leaues fall not off, til they be thrust off, with the knots or buds, wherupon it comes to passe that trees cannot beare fruit plentifully two yeares together, and make themselues ready to blossome against the seasonablenesse of the next Spring.

And if any frost be so extreme, that it stay the sap too much, or too long, then it kils the forward fruit in the bud, and sometimes the tender leaues and twigs, but not the tree. Wherefore, to returne, it is perillous to stop the sap. And where, or when, did you euer see a great tree packt on a wall? Nay, who did euer know a tree so vnkindly splat, come to age? I haue heard of some, that out of their imaginary cunning, haue planted such trees, on the North side of the wall, to auoide drought, but the heate of the Sunne is as comfortable (which they should haue regarded) as the drought is hurtfull. And although water is a soueraigne remedy against drought, ye want of Sun is no way to be helped. Wherefore to conclude this Chapter, let your ground lie so, that it may haue the benefit of the South, and West Sun, and so low and close, that it may haue moysture, and increase his fatnesse (for trees are the greatest suckers & pillers of earth) and (as much as may be) free from great winds.

Chap. 4.
Of the quantity.

It would be remembred what a benefit riseth, not onely to euery particular owner of an Orchard, but also to the common wealth, by fruit, as shall be shewed in the 16. Chapter (God willing) where-upon must needes follow: the greater the Orchard is (being good and well kept) the better it is, for of good things, being equally good, the biggest is the best. Orchard as good as a corn-field. And if it shall appeare, that no ground a man occupieth (no, not the corne field) yeeldeth more gaine to the purse, and house keeping (not to speake of the vnspeakeable pleasure) quantity for quantity, than a good Orchard (besides the cost in planting, and dressing an orchard, is not so much by farre, as the labour and feeding of your corne fields, nor for durance of time, comparable, besides the certainty of the on before the other) I see not how any labour, or cost in this kind, can be idly or wastfully bestowed, or thought too much. Compared with a vinyard. And what other things is a vineyard, in those countries where vines doe thriue, than a large Orchard of trees bearing fruit? Or what difference is there in the iuice of the Grape, and our Cyder & Perry, but the goodnes of the soile & clime where they grow? which maketh the one more ripe, & so more pleasant then the other. What soeuer can be said for the benefit rising from an orchard, that makes for the largenesse of the Orchards bounds. Compared with a garden. And (me thinkes) they do preposterously, that bestow more cost and labours, and more ground in and vpon a garden than vpon an orchard, whence they reape and may reape both more pleasure and more profit, by infinite degrees. And further, that a Garden neuer so fresh, and faire, and well kept, cannot continue without both renewing of the earth and the hearbs often, in the short and ordinary age of a man: whereas your Orchard well kept shall dure diuers hundred yeares, as shall be shewed chap. 14. In a large orchard there is much labour saued, in fencing, and otherwise: for three little orchards, or few trees, being, in a manner, all out-sides, are so blasted and dangered, and commonly in keeping neglected, and require a great fence;

whereas in a great Orchard, trees are a mutuall fence one to another, and the keeping is regarded, and lesse fencing serues sixe acres together, than three in seuerall inclosures.

What quantity of ground. Now what quantity of ground is meetest for an Orchard can no man prescribe, but that must be left to euery mans seuerall iudgement, to be measured according to his ability and will, for other necessaries besides fruite must be had, and some are more delighted with orchard then others.

Want is no hinderance. Let no man hauing a fit plot plead pouerty in this case, for an orchard once planted will maintaine it selfe, and yeeld infinite profit besides. And I am perswaded, that if men did know the right and best way of planting, dressing, and keeping trees, and felt the profit and pleasure thereof, both they that haue no orchards would haue them, & they that haue orchards, would haue them larger, yea fruit-trees in their hedges, as in *Worcester-shire*, &c. How Land-lords by their Tenants may make flourishing Orchards in *England*. And I think, that the want of planting, is a great losse to our common-wealth, & in particular, to the owners of Lord-ships, which Land lords themselues might easily amend, by granting longer terme, and better assurance to their tenants, who haue taken vp this Prouerbe *Botch and sit, Build and flit*: for who will build or plant for an other mans profit? Or the Parliament mighte ioyne euery occupier of grounds to plant and maintaine for so many acres of fruitfull ground, so many seuerall trees or kinds of trees for fruit. Thus much for quantity.

Chap. 5.
Of the forme.

The goodnesse of the soile, and site, are necessary to the wel being of an orchard simply, but the forme is so farre necessary, as the owner shall thinke meete, for that kind of forme wherewith euery particular man is delighted, we leaue it to himselfe, *Suum cuique pulchrum*. The vsuall forme is a square. The forme that men like in generall is a square, for although roundnesse be *forma perfectissima*, yet that principle is good where necessity by art doth not force some other forme. If within one large square the Gardner shall make one round Labyrinth or Maze with some kind of Berries, it will grace your forme, so there be sufficient roomth left for walkes, so will foure or more round knots do. For it is to be noted, that the eye must be pleased with the forme. I haue seene squares rising by degrees with stayes from your house-ward, according to this forme which I haue, *Crassa quod aiunt Minerua,* with an vnsteady hand, rough hewen, for in forming the country gardens, the better sort may vse better formes, and more costly worke. What is needefull more to be sayd, I referre that all (concerning the Forme,) to the Chapter 17 of the ornaments of an Orchard.

A. Al these squares must bee set with trees, the Gardens and other ornaments must stand in spaces betwixt the trees, & in the borders & fences.

B. Trees 20. yards asunder.

C. Garden Knots.

D. Kitchen garden.

E. Bridge.

F. Conduit.

G. Staires.

H. Walkes set with great wood thicke.

I. Walkes set with great wood round about your Or-

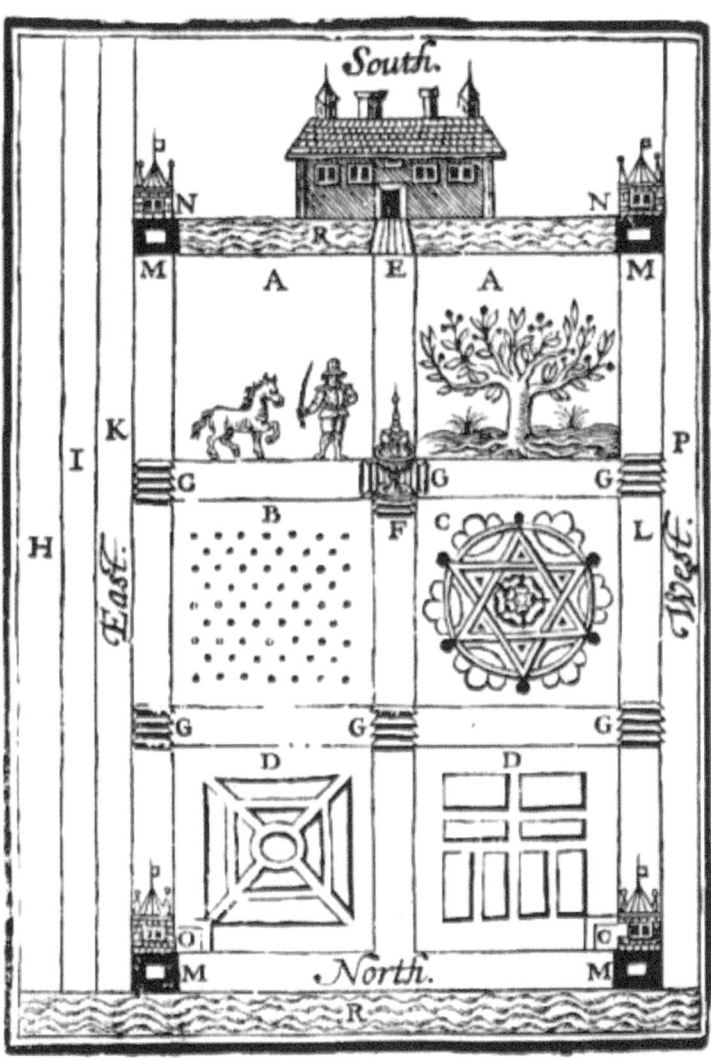

chard.

K. The out fence.

L. The out fence set with stone-fruite.

M. Mount. To force earth for a mount, or such like set it round with quicke, and lay boughes of trees strangely intermingled tops inward, with the earth in the midle.

N. Still-house.

O. Good standing for Bees, if you haue an house.

P. If the riuer run by your doore, & vnder your mount, it will be pleasant.

Chap. 6.
Of Fences.

Effects of euill fencing. All your labour past and to come about an Orchard is lost vnlesse you fence well. It shall grieue you much to see your young sets rubd loose at the rootes, the barke pild, the boughes and twigs cropt, your fruite stolne, your trees broken, and your many yeares labours and hopes destroyed, for want of fences. A chiefe care must be had in this point. You must therefore plant in such a soile, where you may prouide a conuenient, strong and seemely fence. For you can possesse no goods, that haue so many enemies as an orchard, looke Chapter 13. Fruits are so delightsome, and desired of so many (nay, in a manner of all) and yet few will be at cost and take paines to prouide them. Fence well therefore, let your plot be wholly in your owne power, that you make all your fence your selfe: for neighbours fencing is none at all, or very carelesse. Let the fence be your owne. Take heed of a doore or window, (yea of a wall) of any other mans into your orchard: yea, though it be nayld vp, or the wall be high, for   perhaps they will proue theeues.

Kinds of fences, earthen walles. All Fences commonly are made of Earth, Stone, Bricke, Wood, or both earth and wood. Dry wall of earth, and dry Ditches, are the worst fences saue pales or railes, and doe waste the soonest, vnlesse they be well copt with glooe and morter, whereon at *Mighill-tide* it will be good to sow Wall-flowers, commonly called Bee-flowers, or winter Gilly-flowers, because they will grow (though amongst stones) and abide the strongest frost and drought, continually greene and flowring euen in Winter, and haue a pleasant smell, and are timely, (that is, they will floure the first and last of flowers) and are good for Bees. And your earthen wall is good for Bees dry and warme. But these fences are both vnseemly, euill to repaire, and onely for need, where stone or wood cannot be had. Whosoeuer makes such Walles, must not pill the ground in the Orchard, for getting earth, nor make any pits or hallowes, which are both vnseemly and vnprofitable. Old dry earth mixt with sand is best for these. This kind of wall will soone decay, by reason of the trees which grow neere it, for the roots and boales of great trees, will increase, vndermine, and ouerturne such walles, though they were of stone, as is apparant by Ashes, Rountrees, Burt-trees, and such like, carried in the chat, or berry, by birds into stone-walles.

Pale and Raile. Fences of dead wood, as pales, will not last, neither will railes either last or make good fence.

Stone walls. Stone walles (where stone may be had) are the best of this sort, both for fencing, lasting, and shrouding of your young trees. But about this must you bestow much paines and more cost, to haue them handsome, high and durable.

Quicke wood and Moates. But of all other (in mine owne opinion) Quickwood, and Moats or Ditches of water, where the ground is leuell, is the best fence. In vnequall grounds, which will not keepe water, there a double ditch may be cast, made streight and leuel on the top, two yards broad for a faire walke, fiue or sixe foot higher then the soyle, with a gutter on either side, two yards wide, and foure foot deepe set with out, with three or foure chesse of Thorns, and within with Cherry, Plumme, Damson, Bullys, Filbirds, (for I loue these trees better for their fruit, and as well for their forme, as priuit) for you may make them take any forme. And in euery corner (and middle if you will) a mount would be raised, whereabout the wood may claspe, powdered with wood-binde: which wil make with dressing a faire, plesant, profitable, & sure fence. But you must be sure that your quicke thornes either grow wholly, or that there be a supply betime, either with planting new, or plashing the old where need is. And assure your selfe, that neither wood, stone, earth, nor water, can make so strong a fence, as this after seuen yeares growth.

Moates. Moates, Fish-ponds, and (especially at one side a Riuer) within and without your fence, will afford you fish, fence, and moysture to your trees, and pleasure also, if they be so great and deepe that you may haue Swans, & other water birds, good for deuouring of vermine, and boat for many good vses.

It shall hardly auaile you to make any fence for your Orchard, if you be a niggard of your fruit. For as liberality will saue it best from noysome neighbours, liberality I say is the best fence, so Iustice must restraine rioters. Thus when your ground is tempered, squared, and fenced, it is time to prouide for planting.

Chap. 7.
Of Sets.

There is not one point (in my opinion) about an Orchard more to be regarded, than the choyce getting and setting of good plants, either for readinesse or hauing good fruite, or for continuall lasting. For whosoeuer shall faile in the choyce of good Sets, or in getting, or gathering, or setting his plants, shall neuer haue a good or lasting Orchard. And I take want of skill in this faculty to be a chiefe hinderance to the most Orchards, and to many for hauing of Orchards at all.

Slips. Some for readinesse vse slips, which seldome take roote: and if they doe take, they cannot last, both because their roote hauing a maine wound will in short time decay the body of the tree: and besides that rootes being so weakely put, are soone nipt with drought or frost. I could neuer see (lightly) any slip but of apples onely set for trees.

Bur-knot. A Bur-knot kindly taken from an Apple tree, is much better and surer. You must cut him close at the roote ende, an handfull vnder the knot. (Some vse in Summer about *Lammas* to circumcise him, and put earth to the knots with hay roaps, and in winter cut him off and set him, but this is curiosity, needlesse, and danger with remouing, and drought,) and cut away all his twigs saue one, the most principall, which in setting you must leaue aboue the earth, burying his trunk in the crust of the earth for his root. It matters not much what part of the bough the twig growes out of. If it grow out of   or neere the roote end, some say such an Apple will haue no coare nor kirnell. Or if it please the Plantor, he may let his bough be crooked, and leaue out his top end, one foote or somewhat more, wherein will be good grafting, if either you like not, or doubt the fruite of the bough (for commonly your bur-knots are summer fruit) or if you thinke he will not couer his wound safely.

Vsuall Sets. The most vsuall kind of sets, is plants with rootes growing of kirnels of Apples, Peares, and Crabbes, or stones of Cherries, Plummes, &c. Remoued out of a Nursery, Wood or other

Orchard, into, and set in your Orchard in their due places I grant this kind to be better than either of the former, by much, as more sure and more durable. Herein you must note that in sets so remoued, you get all the roots you can; and without brusing of any; Maine rootes cut. I vtterly dislike the opinion of those great Gardners, that following their Bookes would haue the maine rootes cut away, for tops cannot growe without rootes. Stow sets remoued. And because none can get all the rootes, and remouall is an hinderance, you may not leaue on all tops, when you set them: For there is a proportion betwixt the top and root of a tree, euen in the number (at least) in the growth. If the roots be many, they will bring you many tops, if they be not hindred. And if you vse to stow or top your tree too much or too low, and leaue no issue, or little for sap, (as is to be seene in your hedges) it will hinder the growth of rootes and boale, because such a kind of stowing is a kind of smothering, or choaking the sap. Great wood, as Oke, Elme, Ash, &c. being continually kept downe with sheeres, knife, axe, &c. neither boale nor roote will thriue, but as an hedge or bush. If you intend to graff in your Set, you may cut him closer with a greater wound, and nearer the earth, within a foote or two, because the graft or grafts will couer his wound. If you like his fruite, and would haue him to be a tree of himselfe, be not so bold: this I can tell you, that though you do cut his top close, and leaue nothing but his bulke, because his rootes are few, if he be (but little) bigger than your thumbe (as I with all plants remoued to be) he will safely recouer wound within seuen yeares; by good guidance that is. In the next time of dressing immediatly aboue his vppermost sprig, you cut him off aslope cleanely, to that the sprigge stand on the backe side, (and if you can Northward, that the wound may haue the benefit of Sunne) at the vpper ende of the wound: and let that sprigge onely be the boale. Generall rule. And take this for a generall rule; Euery young plant, if he thriue, will recouer any wound aboue the earth, by good dressing, although it be to the one halfe, and to his very heart. Tying of trees. This short cutting at the remoue, saues your plants from Wind, and neede the lesse or no staking. I commend not Lying or Leaning of trees against holds or stayres; for it breedes obstruction of sap and wounds incureable. Generall rule. All remouing of trees as great as your arme, or aboue, is dangerous: though sometime some such will grow but not continue long: Because they be tainted with deadly wounds, either in the roote or top. (And a tree once throughly tainted is neuer good) And though they get some hold in the earth with some lesser taw, or tawes, which giue some nourishment to the body of the tree: yet the heart being tainted, he will hardly euer thriue; Signes of diseases, Chap 13. which you may easily discerne by the blackenesse of the boughes at the heart, when you dresse your trees. Also, when he is set with moe tops than the rootes can nourish, the tops decaying, blacken the boughes, and the boughs the armes, and so they boile at the very heart. Or this taint in the remouall, if it kill not presently, but after some short time, it may be discerned by blacknesse or yellownesse in the barke, and a small hungred leafe. Or if your remoued plant put forth leaues the next and second summer, and little or few spraies, it is a great signe of a taint, and next yeares death. I haue knowne a tree tainted in setting, yet grow, & beare blossomes for diuers yeares: and yet for want of strength could neuer shape his fruit.

Suckers good sets. Next vnto this or rather equall with these plants, are suckers growing out of the roots of great trees, which cherries and plums do seldome or neuer want: and being taken kindly with their roots, will make very good sets. And you may helpe them much by enlarging their rootes with the taws of the tree, whence you take them. They are of two sorts: Either growing from the very root of the tree: and here you must be carefull, not to hurt your tree when you gather them, by ripping amongst the rootes; and that you take them cleane away: for these are a great and continuall annoyance to the growth of your tree: and they will hardly be cleansed. Secondly, or they do arise from some taw: and these may be taken without danger, with long and good rootes, and will soone become trees of strength.

A running Plant. There is another way, which I haue not throughly proued, to get not onely plants for graffing, but sets to remaine for trees, which I call a *Running Plant*: the manner of it is this: Take a roote or kirnell, and put it into the middle of your plot, and the second yeare in the spring, geld his top, if he haue one principall (as commonly by nature they haue) and let him put forth onely foure Cyons toward the foure corners of the orchard, as neere the earth as you can. If he put not foure, (which is rare) stay his top till he haue put so many. When you haue such foure, cut the stocke aslope, as is aforesayd in this chapter, hard aboue the vttermost sprig, & keepe those foure without Cyons cleane and straight, till you haue them a yard and a halfe, at least, or two yards long. Then the next spring in grassing time, lay downe those foure sprayes, towards the foure corners of your Orchard, with their tops in an heape of pure and good earth, and railed as high as the roote of your Cyon (for sap will not descend) and a sod to keepe them downe, leauing nine or twelue inches of the top to looke vpward. In that hill he will put rootes, and his top new Cyons, which you must spread as before, and so from hill to hill till he spread the compasse of your ground, or as farre as you list. If in bending, the Cyons cracke, the matter is small, cleanse the ground and he will recouer. Euery bended bough will put forth branches, and become trees. If this plant be of a burre knot, there is no doubt. I haue proued it in one branch my selfe: and I know at *Wilton* in *Cleeue-land* a Peare-tree of a great bulke and age, blowne close to the earth, hath put at euery knot rootes into the earth, and from roote to top, a great number of mighty armes or trees, filling a great roomth, like many trees, or a little Orchard. Much better may it be done by Art in a lesse tree. And I could not mislike this kind, saue that the time will be long before it come to perfection.

Sets bought. Many vse to buy sets already grafted, which is not the best way: for first, All remoues are dangerous: Againe, there is danger in the carriage: Thirdly, it is a costly course of planting: Fourthly, euery Gardner is not trusty to sell you good fruite: Fifthly, you know not which is best, which is worst, and so may take most care about   your worst trees. Lastly, this way keepes you from practise, and so from experience in so good, Gentlemanly, Scholer-like, and profitable a faculty.

The best sets. The onely best way (in my opinion) to haue sure and lasting sets, is neuer to remoue: for euery remoue is an hinder-ance, if not a dangerous hurt or deadly taint. This is the way. Vnre-moued how. The plot forme being layd, and the plot appointed where you will plant euery set in your orchard, digge the roomth, where your sets shall stand, a yard compasse, and make the earth mellow and cleane, and mingle it with a few coale-ashes, to auoide wormes: and immediately after the first change of the Moone, in the latter end of *February*, the earth being a fresh turn'd ouer, put in euery such roomth three or foure kirnels of Apples or Peares, of the best: euery kirnell in an hole made with your finger, finger deepe, a foote distant one from another: and that day moneth following, as many moe, (lest some of the former misse) in the same compasse; but not in the same holes. Hence (God willing) shall you haue rootes enough. If they all, or diuers of them come vp, you may draw (but not digge) vp (nor put downe) at your pleasure, the next *Nouember*. How many soeuer you take away, to giue or bestow elsewhere, be sure to leaue two of the proudest. And when in your 2. and 3. yeare you Graffe, if you graffe then at all, leaue the one of those two vngraffed, lest in graffing the other you faile: For I find by tryall, that after first or second graffing in the same stocke, being mist (for who hits all) the third misse puts your stocke in deadly danger, for want of issue of sap. Yea, though you hit in graffing, yet may your graffes with winde or otherwise be broken downe. If your graffes or graffe prosper, you haue your desire, in a plant vnremoued,   without taint, and the fruite at your owne choyce, and so you may (some little earth being remooued) pull, but not digge vp the other Plant or Plants in that roomth. If your graffe or stocke, or both perish, you haue another in the same place, of better strength to worke vpon. For thriuing without snub he will ouer-lay your grafted stocke much. And it is hardly possible to misse in graffing so often, if your Gardiner be worth his name.

Sets vngrafted best of all. It shall not be amisse (as I iudge it) if your Kirnels be of choyce fruite, and that you see them come forward proudly in their body, and beare a faire and broad leafe in colour, tending to a greenish yellow (which argues pleasant and great fruit) to try some of them vngraffed: for although it be a long time ere this come to beare fruit, ten or twelue yeares, or moe; and at their first bearing, the fruit will not seeme to be like his owne kind: yet am I assured, vpon tryall, before twenty yeares growth, such trees will increase the bignesse and goodnesse of their fruite, and come perfectly to their owne kind. Trees (like other breeding creatures) as they grow in yeares, bignes and strength, so they mend their fruit. Husbands and Houswiues find this true by experience, in the rearing of their yong store. More then this, there is no tree like this for soundnes and dureable last, if his keeping and dressing be answerable. I grant, the readiest way to come soone to fruit is graffing: because in a manner, all your graffes are taken of fruit bearing trees.

Time of remouing. Now when you haue made choise of your sets to remoue, the ground being ready, the best time is, immediatly after the fall of the leafe, in, or about the change of the Moone, when the sap is most quiet: for then the sap is in turning: for it makes no stay, but in the *extremity* of drought or cold. Generall rule. At any time in winter, may you transplant trees so you put no ice nor snow to the root of your plant in the setting: and therefore open, calme and moist weather is best. To remoue, the leafe being ready to fall and not fallen, or buds apparantly put forth in a moist warme season, for need, sometime may do well: but the safest is to walke in the plaine trodden path.

Some hold opinion that it is best remouing before the fall of the leafe, and I heare it commonly practised in the South by our best arborists, the leafe not fallen: and they giue the reason to be, that the descending of the sap will make speedy rootes. But marke the reasons following and I thinke you shall find no soundnesse, either in that position or practise, at least in the reason.

1. I say, it is dangerous to remoue when the sap is not quiet, for euery remoue giues a maine checke to the stirring sap, by staying the course therof in the body of your plant, as may appeare in trees remoued any time in summer, they commonly dye, nay hardly shall you saue the life of the most young and tender plant of any kinde of wood (scarcely herbes) if you remoue them in the pride of sap. For proud sap vniuersally staied by remoual, euer hinders; often taints and so presently, or in very short time kills. Sap is like bloud in mans body, in which is the life, *Cap. 3. p. 9.* If the blood vniuersally be cold, life is excluded; so is sap tainted by vntimely remouall. A stay by drought, or cold, is not so dangerous (though dangerous if it be extreme) because more naturall.

2. The sap neuer descends, as men suppose, but is consollidated & transubstantiated into the substance of the tree, and passeth (alwayes aboue the earth) vpward, not onely betwixt the barke and the wood, but also into and in both body & barke, though not so plentifully, as may appeare by a tree budding, nay fructifying two or

three yeres, after he be circumcised at the very root, like a riuer that inlargeth his channel by a continual descent.

3. I cannot perceiue what time they would haue the sap to descend. At *Midsommer* in a biting drought it staies, but descends not, for immediatly vpon moisture it makes second shoots, at (or before rather) *Michaeltide*, when it shapens his buds for next yeares fruit. If at the fal of leafe, I grant, about that time is the greatest stand, but no descent, of sap, which begins somwhat before the leafe fall, but not long, therfore at that time must be the best remouing, not by reason of descent, but stay of sap.

4. The sap in this course hath his profitable and apparant effects, as the growth of the tree, couering of wounds, putting of buds, &c. Wherupon it follows, if the sap descend, it must needs haue some effect to shew it.

5. Lastly, boughs plasht and laid lower then the root, dye for want of sap descending, except where it is forced by the maine streame of the sap, as in top boughs hanging like water in pipes, or except the plasht bough lying on the ground put rootes of his owne, yea vnder boughs which we commonly call water boughs, can scarcely get sap to liue, yea in time dye, because the sap doth presse so violently vpward, and therefore the fairest shootes and fruits are always in the top.

Remooue soone. *Obiect.* If you say that many so remoued thriue, I say that somewhat before the fall of the leafe (but not much) is the stand, for the fall & the stand are not at one instant, before the stand is dangerous. But to returne.

The sooner in winter you remoue your sets, the better; the latter the worse: For it is very perillous if a strong drought take your Sets before they haue made good their rooting. A Plant set at the fall, shall gaine (in a a manner) a whole yeeres growth of that which is set in the Spring after.

The manner of setting. I vse in the setting to be sure, that the earth be mouldy, (and somewhat moist) that it may runne among the small tangles without straining or bruising: and as I fill in earth to his root, I shake the Set easily to and fro, to make the earth settle the better to his roots: and withall easily with my foot I put in the earth close; for ayre is noysome, and will follow concauities. Some prescribe Oates to be put in with the earth. I could like it, if I could know any reason thereof: and they vse to set their Plant with the same side toward the Sunne: but this conceit is like the other. For first I would haue euery tree to stand so free from shade, that not onely the root (which therefore you must keepe bare from graffe) but body, boughes, and branches, and euery spray, may haue the benefit of Sunne. And what hurt, if that part of the tree, that before was shadowed, be now made partaker of the heat of the Sunne? In turning of Bees, I know it is hurtfull, because it changeth their entrance, passage, and whole worke: But not so in Trees.

Set in the crust. Set as deepe as you can, so that in any wise you goe not beneath the crust. Looke Chap. 2.

Moysture good. We speake in the second Chapter of moysture in generall: but now especially hauing put your remoued plant into the earth, powre on water (of a puddle were good) by distilling presently, and so euery weeke twice in strong drought, so long as the earth will drinke, and refuse by ouerflowing. For moisture mollifies, and both giues leaue to the roots to spread, and makes the earth yeeld sap and nourishment with plenty & facility. Nurses (they say) giue most & best milke after warme drinks.

If your ground be such that it will keepe no moisture at the root of your plant, such plant shall neuer like, or but for a time. There is nothing more hurtfull for young trees then piercing drought. I haue known trees of good stature after they haue beene of diuers yeeres growth, & thriue well for a good time, perish for want of water, and very many by reason of taints in setting.

It is meet your sets and grafts be fenced, till they be as big as your arme, for feare of annoyances. Grafts must be fenced. Many waies may sets receiue dammages, after they be set, whether grafted or vngrafted. For although we suppose, that no noysome beast, or other thing must haue accesse among your trees: yet by casualty, a Dog, Cat, or such like, or your selfe, or negligent friend bearing you company, or a shrewd boy, may tread or fall vpon a young and tender plant or graft. To auoid these and many such chances, you must stake them round a pretty distance from the set, neither so neere, nor so thicke, but that it may haue the benefit of Sun, raine, and ayre. Your stakes (small or great) would be so surely put, or driuen into the earth, that they breake not, if any thing happen to leane vpon them, else may the fall be more hurtfull, then the want of the fence. Let not your stakes shelter any weeds about your sets, for want of Sunne is a great hinderance. Let them stand so farre off, that your grafts spreading receiue no hurt, either by rubbing on them, or of any other thing passing by. If your stocke be long, and high grafted (which I must discommend (except in need) because there the sap is weake, and they are subiect to strong wind, and the lighting of birds) tie easily with a soft list three or foure prickes vnder the clay, and let their tops stand aboue the grafts, to auoid the lighting of Crowes, Pyes, &c. vpon your grafts. If you sticke some sharpe thornes at the roots of your stakes, they will make hurtfull things keepe off the better. Other better fences for your grafts I know none. And thus much for sets and setting.

Chap. 8.
Of the distance of Trees.

I Know not to what end you should prouide good ground, well fenced, & plant good sets; and when your trees should come to profit, haue all your labours lost, for want of due regard to the distance of placing your trees. I haue seene many trees stand so thicke, that one could not thriue for the throng of his neighbours. Hurts of too neere planting. If you doe marke it, you shall see the tops of trees rubd off, their sides galled like a galled horses backe, and many trees haue more stumps then boughes, and most trees no well thriuing, but short, stumpish, and euill thriuing boughes: like a Corne field ouer seeded, or a towne ouer peopled, or a pasture ouer-laid, which the Gardiner must either let grow, or leaue the tree very few boughes to beare fruit. Hence small thrift, galls, wounds, diseases, and short life to the trees: and while they liue greene, little, hard, worme-eaten, and euill thriuing fruit arise, to the discomfort of the owners.

Remedy. To preuent which discommodity, one of the best remedies is the sufficient and fit distance of trees. Therefore at the setting of your plants you must haue such respect, that the distance of them be such, that euery tree be not annoyance, but an helpe to his fellowes: for trees (as all other things of the same kind) should shroud, and   not hurt one another. And assure your selfe that euery touch of trees (as well vnder as aboue the earth) is hurtfull. Generall rule. All touches hurtfull. Therefore this must be a generall rule in this Art: That no tree in an Orchard well ordered, nor bough, nor Cyon, drop vpon, or touch his fellowes. Let no man thinke this vnpossible, but looke in the eleuenth Chapter of dressing of trees. If they touch, the winde will cause a forcible rub. Young twigs are tender, if boughes or armes touch or rub, if they are strong, they make great galls. No kind of touch therefore in trees can be good.

The best distance of trees. Now it is to be considered what distance amongst sets is requisite, and that must be gathered from the compasse and roomth, that each tree by probability will take and fill. And herein I am of a contrary opinion to all them, which practise or teach the planting of trees, that euer yet I knew, read, or heard of. For the common space betweene tree and tree is ten foot: if twenty foot, it is thought very much. But I suppose twenty yards distance is small enough betwixt tree and tree, or rather too too little. For the distance must needs be as far as two trees are well able to ouer spread, and fill, so they touch not by one yard at least. Now I am assured, and I know one Apple-tree, set of a slip *finger-great*, in the space of 20 yeares, (which I account a very small part of a trees age, as is shewed Chapter 14.) hath spred his boughes eleuen or twelue yards compasse, that is, fiue or six yards on euery side. Here I gather, that in forty or fifty yeares (which yet is but a small time of his age) a tree in good soile, well liking, by good dressing (for that is much auaileable to this purpose) will spread double at the least, viz. twelue yards on a side, which being added to twelue alotted to his fellow, make twenty and foure yards, and   so farre distant must euery tree stand from another. And looke how farre a tree spreads his boughes aboue, so far doth he put his roots vnder the earth, or rather further, if there be no stop, nor let by walls, trees, rocks, barren earth and such like: for an huge bulk, and strong armes, massie boughes, many branches, and infinite twigs, require wide spreading roots. The parts of a tree. The top hath the vast aire to spread his boughs in, high and low, this way and that way: but the roots are kept in the crust of the earth, they may not goe downward, nor vpward out of the earth, which is their element, no more then the fish out of the water, Camelion out of the Aire, nor Salamander out of the fire. Therefore they must needs spread farre vnder the earth. And I dare well say, if nature would giue leaue to man by Art, to dresse the roots of trees, to take away the tawes and tangles, that lap and fret and grow superfluously and disorderly, (for euery thing *sublunary* is cursed for mans sake) the tops aboue being answerably dressed, we should haue trees of wonderfull greatnes, and infinite durance. And I perswade myselfe that this might be done sometimes in Winter, to trees standing in faire plaines and kindly earth, with small or no danger at all. So that I conclude, that twenty foure yards are the least space that Art can allot for trees to stand distant one from another.

Waste ground in an Orchard. If you aske me what vse shall be made of that waste ground betwixt tree and tree? I answer: If you please to plant some tree or trees in that middle space, you may, and as your trees grow contigious, great and thick, you may at your pleasure take vp those last trees. And this I take to be the chiefe cause, why the most trees stand so thicke. For men not knowing (or not regarding) this secret of needfull distance, and louing fruit of trees planted to their handes, thinke much to pull vpp any, though they pine one another. If you or your heires or successors would take vp some great trees (past setting) where they stand too thicke, be sure you doe it about *Midsummer*, and leaue no maine root. I destinate this space of foure and twenty yards, for trees of age & stature. More then this, you haue borders to be made for walkes with Roses, Berries, &c.

And chiefly consider: that your Orchard, for the first twenty or thirty yeeres, will serue you for many Gardens, for Safron, Licoras, roots, and other herbs for profit, and flowers for pleasure: so that no ground need be wasted if the Gardiner be skillfull and diligent. But be sure you come not neere with such deepe deluing the roots of your trees, whose compasse you may partly discerne, by the compasse of the tops, if your top be well spread. And vnder the droppings and shadow of your trees, be sure no herbes will like. Let this be said for the distance of Trees.

Chap. 9.
Of the placing of Trees.

The placing of trees in an Orchard is well worth the regard: For although it must be granted, that any of our foresaid trees (Chap. 2.) will like well in any part of your Orchard, being good and well drest earth: yet are not all Trees alike worthy of a good place. And therefore I wish that your Filbird, Plummes, Damsons, Bulesse, and such like, be vtterly remoued from the plaine soile of your Orchard into your fence: for there is not such fertility and easefull growth, as within: and there also they are more subiect to, and can abide the blasts of *Æolus*. The cherries and plummes being ripe in the hot time of Summer, and the rest standing longer, are not so soone shaken as your better fruit: neither if they suffer losse, is your losse so great. Besides that, your fences and ditches will deuoure some of your fruit growing in or neere your hedges. And seeing the continuance of all these (except Nuts) is small, the care of them ought to be the lesse. And make no doubt but the fences of a large Orchard will containe a sufficient number of such kind of Fruit trees in the whole compasse. It is not material, but at your pleasure, in the said fences, you may either intermingle your seuerall kinds of fruit-trees, or set euery kind by himselfe, which order doth very well become your better and greater fruit. Let therefore your Apples, Peares, and Quinches, possesse all the soile of your Orchard, vnlesse you be especially affected to some of your other kinds: and of them let your greatest trees of growth stand furthest from Sunne, and your Quinches at the South side or end, and your Apples in the middle, so shall none be any hinderance to his fellowes. The Warden-tree, and Winter-Peare will challenge the preheminence for stature. Of your Apple-trees you shall finde difference in growth. A good Pippin will grow large, and a Costard-tree: stead them on the North side of your other Apples, thus being placed, the least will giue Sun to the rest, and the greatest will shroud their fellowes. The fences and out-trees will guard all.

Chap. 10.
Of Grafting.

Of Grauing or Caruing.
Grafting What. Now are we come to the most curious point of our

faculty: curious in conceit, but indeede as plaine and easie as the rest, when it is plainely shewne, which we commonly call *Graffing*, or (after some) *Grafting*. I cannot *Etymologize*, nor shew the originall of the Word, except it come of *Grauing* and *Caruing*. A Graffe. But the thing or matter is: The reforming of the fruite of one tree with the fruit of another, by an artificiall transplacing, or transposing of a twigge, bud or leafe, (commonly called a *Graft*) taken from one tree of the same, or some other kind, and placed or put to, or into another tree in one time and manner.

Kinds of grafting. Of this there be diuers kinds, but three or foure now especially in vse: to wit, Grafting, incising, packing on, grafting in the scutchion, or inoculating: whereof the chiefe and most vsuall, is called grafting (by the generall name, *Catahexocen*:) for it is the most knowne, surest, readiest, and plainest way to haue store of good fruit.

Graft how. It is thus wrought: You must with a fine, thin, strong and sharpe Saw, made and armed for that purpose, cut off a foot aboue the ground, or thereabouts, in a plaine without a knot, or as neere as you can without a knot (for some Stocks will be knotty) your Stocke, set, or plant, being surely stayed with your foot and legge, or otherwise straight ouerthwart (for the Stocke may be crooked) and then plaine his wound smoothly with a sharpe knife:

that done, cleaue him cleanly in the middle with a cleauer, and a knocke or mall, and with a wedge of wood, Iron or Bone, two handfull long at least, put into the middle of that clift, with the same knocke, make the wound gape a straw bredth wide, into which you must put your Graffes.

A Graft what. The graft is a top twig taken from some other Tree (for it is folly to put a graffe into his owne Stocke) beneath the vppermost (and sometime in need the second) knot, and with a sharpe knife fitted in the knot (and some time out of the knot when need is) with shoulders an ynch downeward, and so put into the stocke with some thrusting (but not straining) barke to barke inward.

Eyes. Let your graffe haue three or foure eyes, for readinesse to put forth, and giue issue to the sap. It is not amisse to cut off the top of your graffe, and leaue it but fiue or six inches long, because commonly you shall see the tops of long graffes die. The reason is this. The sap in graffing receiues a rebuke, and cannot worke so strongly presently, and your graffes receiue not sap so readily, as the naturall branches. When your graffes are cleanely and closely put in, and your wedge puld out nimbly, for feare of putting your graffes out of frame, take well tempered morter, soundly wrought with chaffe or horse dung (for the dung of cattell will grow hard, and straine your graffes) the quantity of a Gooses egge, and diuide it iust, and therewithall, couer your stocke, laying the one halfe on the one side and the other halfe on the other side of your graffes (for thrusting against your graffes) you moue them, and let both your hands thrust at once, and alike, and let your clay be tender, to yeeld easily; and all, lest you moue your graffes. Some vse to couer the clift of the Stocke, vnder the clay with a piece of barke or leafe, some with a sear-cloth of waxe and butter, which as they be not much needfull, so they hurt not, vnlesse that by being busie about them, you moue your graffes from their places. They vse also mosse tyed on aboue the clay with some bryer, wicker, or other bands. These profit nothing. Generall rule. They all put the graffes in danger, with pulling and thrusting: for I hold this generall rule in graffing and planting: if your stocke and graffes take, and thriue (for some will take and not thriue, being tainted by some meanes in the planting or graffing) they will (without doubt) recouer their wounds safely and shortly.

Time of graffing. The best time of graffing from the time of remouing your stocke is the next Spring, for that saues a second wound, and a second repulse of sap, if your stocke be of sufficient bignesse to take a graffe from as big as your   thumbe, to as big as an arme of a man. You may graffe lesse (which I like) and bigger, which I like not so well. The best time of the yeere is in the last part of *February*, or in *March*, or beginning of *Aprill*, when the Sunne with his heat begins to make the sap stirre more rankely, about the change of Moone before you see any great apparancy of leafe or flowers but onely knots and buds, and before they be proud, though it be sooner. Cheries, Peares, Apricocks, Quinces, and Plummes would be gathered and grafted sooner.

Gathering graffes. The graffes may be gathered sooner in *February*, or any time within a moneth, or two before you graffe or vpon the same day (which I commend) If you get them any time before, for I haue knowne graffes gathered in *December*, and doe well, take heed of drought. I haue my selfe taken a burknot of a tree, & the same day when he was laid in the earth about mid *February*, gathered grafts and put in him, and one of those graffes bore the third yeere after, and the fourth plentifully. Graffes of old trees. Graffes of old trees would be gathered sooner then of young trees, for they sooner breake and bud. If you keepe graffes in the earth, moisture with the heat of the Sun will make them sprout as fast, as if they were growing on the tree. And therefore seeing keeping is dangerous, the surest way (as I iudge) is to take them within a weeke of the time of your grafting.

The grafts would be taken not of the proudest twigs, for it may be your stocke is not answerable in strength. Where taken. And therefore say I, the grafts brought from South to vs in the North although they take and thriue (which is somewhat doubtfull, by reason of the difference of the Clime and carriage) yet shall they in time fashion themselues to our cold Northerne soile, in growth, taste &c.

Nor of the poorest, for want of strength may make them vnready to receiue sap (and who can tell but a poore graft is tainted) nor on the outside of your tree, for there should your tree spread but in the middest; for there you may be sure your Tree is no whit hindered in his growth or forme. He will stil recouer inward, more then you would wish. Emmits. If your clay clift in Summer with drought, looke well in the Chinkes for Emmits and Earewigs, for they are cunning and close theeues about grafts you shall finde them stirring in the morning and euening, and the rather in the moist weather. I haue had many young buds of Graffes, euen in the flourishing, eaten with Ants. Let this suffice for graffing, which is in the faculty counted the chiefe secret, and because it is most vsuall it is best knowne.

Graffes are not to be disliked for growth, till they wither, pine, and die. Vsually before *Midsummer* they breake, if they liue. Some (but few) keeping proud and greene, will not put till the second yeere, so is it to be thought of sets.

The first shew of putting is no sure signe of growth, it is but the sap the graffe brought with him from his tree.

So soone as you see the graft put for growth, take away the clay, for then doth neither the stocke nor the graffe need it (put a little fresh well tempered clay in the hole of the stocke) for the clay is now tender, and rather keepes moistture then drought.

The other waies of changing the naturall fruit of Trees, are more curious then profitable, and therefore I mind not to bestow much labour or time about them, onely I shall make knowne what I haue proued, and what I doe thinke.

Incising. And first of incising, which is the cutting of the backe of the boale, a rine or branch of a tree at some bending or knee, shoulderwise with two gashes, onely with a sharpe knife to the wood: then take a wedge, the bignes of your graffe sharpe ended, flat on the one side, agreeing with the tree, and round on the other side, and with that being thrust in, raise your barke, then put in your graffe, fashioned like your wedge iust: and lastly couer your wound, and fast it vp, and take heed of straining. A great stocke. This will grow but to small purpose, for it is weake hold, and lightly it will be vnder growth. Thus may you graft betwixt the barke and the tree of a great stocke that will not easily be clifted: But I haue tryed a better way for great trees, viz First, cut him off straight, and cleanse him with your knife, then cleaue him into foure quarters, equally with a strong cleauer: then take for euery Clift two or three small (but hard) wedges iust of the bignesse of your grafts, and with those Wedges driuen in with an hammer open the foure clifts so wide (but no wider) that they may take your foure graffes, with thrusting not with straining: and lastly couer and clay it closely, and this is a sure and good way of grafting: or thus, clift your stocke by his edges twice or thrice with your cleauer, and open him with your wedge in euery clift one by one, and put in your grafts, and then couer them. This may doe well.

Packing thus. Packing on is, when you cut aslope a twig of the same bignesse with your graft, either in or besides the knot, two inches long, and make your graft agree iumpe with the Cyon, and gash your graft and your Cyon in the middest of the wound, length-way, a straw breadth deepe, and thrust the one into the other, wound to wound, sap to sap, barke to barke, then tie them close and clay them. This may doe well. The fairest graft I haue in my little Orchard, which I haue planted, is thus packt on, and the branch whereon I put him, is in his plentifull roote.

To be short in this point, cut your graft in any sort or fashion, two inches long, and ioyne him cleanly and close to any other sprig of any tree in the latter end of the time of grafting, when sap is some-what rife, and in all probability they will close and thriue: thus

The Sprig. The graft. The twig. The graft.

Or any other fashion you thinke good.

Inoculating. Inoculating is an eye or bud, taken barke and all from one tree, and placed in the roome of another eie or bud of another, cut both of one compasse, and there bound. This must be done in Summer, when the sap is proud.

Much like vnto this is that, they call grafting in the scutchion, they differ thus: That here you must take an eie with his leafe, or (in mine opinion) a bud with his leaues. Graffing in the Scutchion. (Note that an eie is for a Cyon, a bud is for flowers and fruit,) and place them on another tree, in a plaine (for so they teach) the place or barke where you must set it, must be thus cut H with a sharpe knife, and the barke raised with a wedge, and then the eie or budde put in and so bound vp. I cannot denie but such may grow. And your bud if he take will flowre and beare fruit that yeere: as some grafts & sets also, being set for bloomes. If these two kinds thriue, they reforme but a spray, and an vndergrowth. Thus you may place Roses on Thornes, and Cherries on Apples, and such like. Many write much more of grafting, but to small purpose. Whom we leaue to themselues, & their followers; & ending this secret we come in the next   Chapter to a point of knowledge most requisite in an Arborist, as well for all other woods as for an Orchard.

Chap. 11.
Of the right dressing of Trees.

Necessity of dressing trees. If all these things aforesaid were indeed performed, as we haue shewed them in words, you should haue a perfect Orchard in nature and substance, begunne to your hand; And yet are all these things nothing, if you want that skill to keepe and dresse your trees. Such is the condition of all earthly things, whereby a man receiueth profit or pleasure, that they degenerate presently without good ordering. Man himselfe left to himselfe, growes from his heauenly and spirituall generation, and becommeth beastly, yea deuillish to his owne kind, vnlesse he be regenerate No maruell then, if Trees make their shootes, and put their spraies disorderly. And truly (if I were worthy to iudge) there is not a mischiefe that breedeth greater and more generall harme to all the Orchard (especially if they be of any continuance) that euer I saw, (I will not except three) then the want of the skilfull dressing of trees. It is a common and vnskilfull opinion, and saying. Let all grow, and they will beare more fruit: and if you lop away superfluous boughes, they say, what a pitty is this? Generall rule. How many apples would these haue borne? not considering there may arise hurt to your Orchard, as well (nay rather) by abundance, as by want of wood. Sound and thriuing plants in a good soile, will euer yeeld too much wood, and disorderly, but neuer too little. So that a skilfull and painfull Arborist, need   neuer want matter to effect a plentifull and well drest Orchard: for it is an easie matter to take away superfluous boughes (if your Gardner haue skill to know them) whereof your plants will yeeld abundance, and skill will leaue sufficient well ordered. All ages both by rule and experience doe consent to a pruning and lopping of trees: yet haue not any that I know described vnto vs (except in darke and generall words) what or which are those superfluous boughes, which we must take away, and that is the chiefe and most needfull point to be knowne in lopping. And we may well assure our selues, (as in all other Arts, so in this) there is a vantage and dexterity, by skill, and an habite by practise out of experience, in the performance hereof for the profit of mankind; yet doe I not know (let me speake it with the patience of our cunning Arborists) any thing within the compasse of humane affaires so necessary, and so little regarded, not onely in Orchards, but also in all other timber trees, where or whatsoeuer.

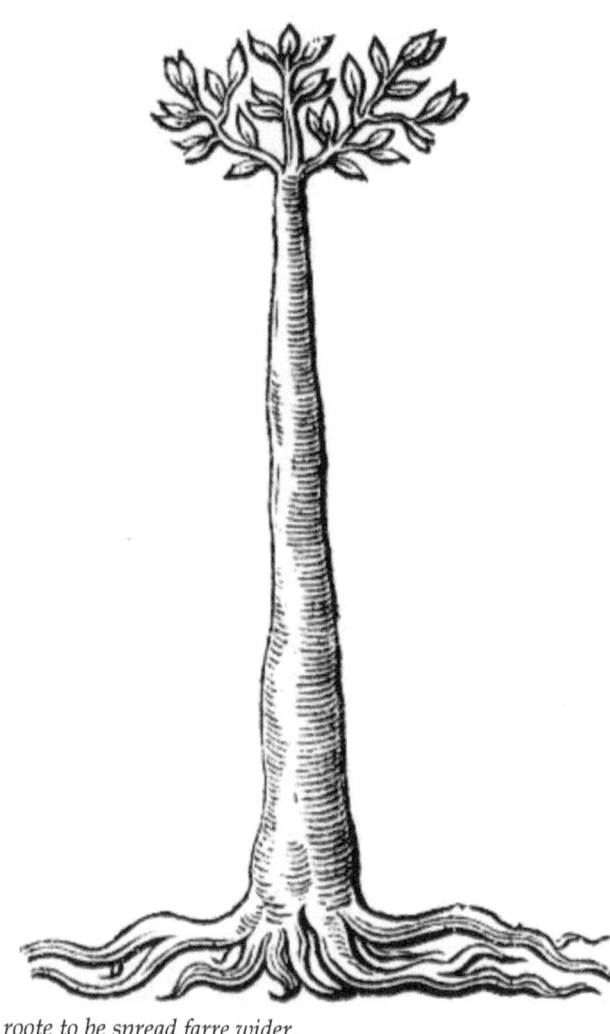

Imagine the roote to be spread farre wider.

Timber wood euill drest. How many forrests and woods? wherein you shall haue for one liuely thriuing tree, foure (nay sometimes 24.) euill thriuing, rotten and dying trees, euen while they liue. And instead of trees thousands of bushes and shrubs. What rottennesse? what hollownesse? what dead armes? withered tops? curtailed trunks? what loads of mosses? drouping boughes? and dying branches shall you see euery where? And those that like in this sort are in a manner all vnprofitable boughes, canked armes, crooked, little and short boales: what an infinite number of bushes, shrubs, and skrogs of hazels, thornes, and other profitable wood, which might be brought by dressing to become great and goodly trees. The cause of hurts in woods. Consider now the cause: The lesser wood hath beene spoiled   with carelesse, vnskilfull, and vntimely stowing, and much also of the great wood. The greater trees at the first rising haue filled and ouer-loaden themselues with a number of wastfull boughes and suckers, which haue not onely drawne the sap from the boale, but also haue made it knotty, and themselues and the boale mossie for want of dressing, whereas if in the prime of growth   they had bene taken away close, Dresse timber trees how. all but one top (according to this patterne) and cleane by the bulke, the strength of all the sap should haue gone to the bulke, and so he would haue recouered and couered his knots, and haue put forth a faire, long and streight body (as you see) for timber profitable, huge great of bulke, and of infinite last.

If all timber trees were such (will some say) how should we haue crooked wood for wheeles, courbs, &c.

Answ. Dresse all you can, and there will be enough crooked for those vses.

More than this, in most places, they grow so thicke, that neither themselues, nor earth, nor any thing vnder or neere them can thriue, nor Sunne, nor raine, nor aire can doe them, nor any thing neere or vnder them any profit or comfort.

I see a number of Hags, where out of one roote you shall see three or foure (nay more, such as mens vnskilfull greedinesse, who desiring many haue none good) pretty Okes or Ashes straight and tall, because the root at the first shoote giues sap amaine: but if one onely of them might bee suffered to grow, and that well and cleanely pruned, all to his very top, what a tree should we haue in time? And we see by those rootes continually and plentifully springing, notwithstanding so deadly wounded. What a commodity should arise to the owner, and the Common-wealth, if wood were cherished, and orderly dressed.

Profit of trees dressed. The wast boughes closely and skilfully taken away, would giue vs store of fences and fewell, and the bulke of the tree in time would grow of huge length and bignes. But here (me thinkes) I heare an vnskilfull Arborist say, that trees haue their seuerall formes, euen by nature, the Peare, the Holly, the Aspe, &c. grow long in bulke with few and little armes, the Oke by nature broad, and such like. All this I graunt: but grant me also, that there is a profitable end, and vse of euery tree, from which if it decline (though by nature) yet man by art may (nay must) correct it. The end of Trees. Now other end of trees I neuer could learne, than good timber, fruit much and good, and pleasure. Vses physicall hinder nothing a good forme.

Trees will take any forme. Neither let any man euer so much as thinke, that it vnprobable, much lesse vnpossible, to reforme any tree of what kind soeuer. For (beleeue me) I haue tried it, I can bring any tree (beginning by time) to any forme. The peare and holly may be made to spread, and the Oke to close.

But why do I wander out of the compasse of mine Orchard, into the Forrests and Woods? Neither yet am I from my purpose, if boales of timber trees stand in need of all the sap, to make them great and straight (for strong growth and dressing makes strong trees) then it must needes be profitable for fruit (a thing more immediately seruing a mans need) to haue all the sap his roote can yeeld: The end of Trees. for as timber sound, great and long, is *the good of timber trees*, and therefore they beare no fruite of worth: so fruit, good, sound, pleasant, great and much, is the end of fruit-trees. That gardner therefore shall performe his duty skilfully and faithfully, which shall so dresse his trees, that they may beare such and such store of fruit, which he shall neuer do (dare vndertake) vnlesse he keepe this order in dressing his trees.

How to dresse a fruit-tree. A fruit tree so standing, that there need none other end of dressing but fruit (not ornaments for walkes, nor delight to such as would please their eye onely, and yet the best forme can not but both adorne and delight) must be parted from within two foote, or thereabouts, of the earth, so high to giue liberty to dresse his roote, and no higher, for drinking vp the sap that should feede his fruit, for the boale will be first, and best serued and fed, because he is next the roote, and of grenest waxe and substance, and that makes him longest of life, into two, three, or foure armes, as your stocke or graffes yeelde twigs, and euery arme into two or more branches, and euery branch into his seuerall Cyons, still spreading by equall degrees, so that his lowest spray be hardly without the reach of a mans hand, and his highest be not past two yards higher, rarely (especially in the middest) that no one twig touch his fellow. Let him spread as farre as he list without his maister-bough or lop equally. And when any bough doth grow sadder and fall lower, than his fellowes (as they will with weight of fruite) ease him the next spring of his superfluous twigs, and he will rise: when any bough or spray shall amount aboue the rest; either snub his top with a nip betwixt your finger and your thumbe, or with a sharpe knife, and take him cleane away, and so you may vse any Cyon you would reforme, and as your tree shall grow in stature and strength, so let him rise with his tops, but slowly, and earely, especially in the middest, and equally, and in bredth also, and follow him vpward with lopping his vndergrowth and water boughes, keeping the same distance of two yards, but not aboue three in any wise, betwixt the lowest and the highest twigs.

Benefits of good dressing. 1. Thus you shall haue well liking, cleane skind, healthfull great, and long-lasting trees.

2. Thus shall your tree grow low, and safe from winds, for his top will be great, broad and weighty.

3. Thus growing broad, shall your trees beare much fruit (I dare say) one as much as six of your common trees, and good without shadowing, dropping and fretting: for his boughes, branches, and twigs shalbe many, and those are they (not the boale) which beare the fruit.

4. Thus shall your boale being little (not small but low) by reason of his shortnesse, take little, and yeeld much sap to the fruit.

5. Thus your trees by reason of strength in time of setting shall put forth more blossomes, and more fruite, being free from taints; for strength is a great helpe to bring forth much and safely, whereas weakenesse failes in setting though the season be calme.

Some vse to bare trees rootes in Winter, to stay the setting til hotter seasons, which I discommend, because,

1. They hurt the rootes.

2. It stayes it nothing at all.

3. Though it did, being small, with vs in the North, they haue their part of our *Aprill* and *Mayes* frosts.

4. Hinderance cannot profit weake trees in setting.

5. They wast much labour.

6. Thus shall your tree be easie to dresse, and without danger, either to the tree or the dresser.

7. Thus may you safely and easily gather your fruite without falling, bruising or breaking of Cyons.

This is the best forme of a fruit tree, which I haue here onely shadowed out for the better capacity of them that are led more with the eye, than the mind, crauing pardon for the deformity, because I am nothing skilfull either in painting or caruing.

Imagine that the paper makes but one side of the tree to appeare, the whole round compasse will giue leaue for many more armes, boughes, branches, and Cyons.

The perfect forme of a Fruit-tree.

If any thinke a tree cannot well be brought to this forme: *Experto crede Roberto*, I can shew diuers of them vnder twenty yeeres of age.

Time best for proining. The fittest time of the Moone for proyning is as of grafting, when the sap is ready to stirre (not proudly stirring) and so to couer the wound, and of the yeere, a moneth before (or at least when) you graffe. Dresse Peares, Apricocks, Peaches, Cherries, and Bullys sooner. And old trees before young plants, you may dresse at any time betwixt Leafe and Leafe. And note, where you take any thing away, the sap the next Summer will be putting: be sure therefore when he puts a bud in any place where you would not haue him, rub it off with your finger.

Dressing betime. And here you must remember the common homely Prouerbe:

Soone crookes the Tree,
That good Camrell must be.

Beginne betime with trees, and do what you list: but if you let them grow great and stubborne, you must do as the trees list. They will not bend but breake, nor bee wound without danger. A small branch will become a bough, and a bough an arme in bignesse. Then if you cut him, his wound will fester, and hardly, without good skill, recouer: therefore, *Obsta principys*. Faults of euill drest trees, and the remedy. Of such wounds, and lesser, of any bough cut off a handfull or more from the body, comes hollownesse, and vntimely death. And therefore when you cut, strik close, and cleane, and vpward, and leaue no bunch.

The forme altered. This forme in some cases sometimes may be altered: If your tree, or trees, stand neere your Walkes, if it please your fancy more, let him not breake, till his boale be aboue you head: so may you walke vnder your trees at your pleasure. Or if you set your fruit-trees for your shades in your Groues, then I expect not the forme of the tree, but the comelinesse of the walke.

Dressing of old trees. All this hitherto spoken of dressing, must be vnderstood of young plants, to be formed: it is meete somewhat be sayd for the instruction of them that haue olde trees already formed, or rather deformed: for, *Malum non vitatur nisi cognitum*. The faults therefore of the disordered tree, I find to be fiue:

Faults are fiue, and their remedies. 1. An vnprofitable boale.
2. Water-boughes.
3. Fretters.
4. Suckers: And,
5. One principall top.

1. Long boale. A long boale asketh much feeding, and the more he hath the more he desires, and gets (as a drunken man drinke, or a couetuous man wealth) and the lesse remaines for the fruit, he puts his boughes into the aire, and makes them, the fruit, and it selfe more dangered with windes: for this I know no remedy, after that the tree is come to growth, once euill, neuer good.No remedy.

2. Water boughs. Water boughes, or vndergrowth, are such boughes as grow low vnder others and are by them ouergrowne, ouershadowed, dropped on, and pinde for want of plenty of sap, and by that meanes in time die: For the sap presseth vpward; and it is like water in her course, where it findeth most issue, thither it floweth, leauing the other lesser floes dry: euen as wealth to wealth,

and much to more. These so long as they beare, they beare lesse, worse, and fewer fruit, and waterish.

Remedy. The remedy is easie if they be not growne greater then your arme. Lop them close and cleane, and couer the midel of the wound, the next Summer when he is dry, with a salue made of tallow, tarre, and a very little pitch, good for the couering of any such wound of a great tree: Barke-pild, and the remedy. vnlesse it be barke-pild, and then sear-cloath of fresh Butter, Hony, and Waxe, presently (while the wound is greene) applyed, is a soueraigne remedy in Summer especially. Some bind such wounds with a thumbe rope of Hay, moist, and rub it with dung.

Fretters. Fretters are, when as by the negligence of the Gardner, two or moe parts of the tree, or of diuers trees, as armes, boughes, branches, or twigs, grow to neere and close together, that one of them by rubbing, doth wound another. Touching. This fault of all other shewes the want of skill or care (at least) in the Arborist: for here the hurt is apparant, and the remedy easie, seene to betime: galls and wounds incurable, but by taking away those members: Remedy. for let them grow, and they will be worse and worse, & so kill themselues with ciuill strife for roomth, and danger the whole tree. Auoide them betime therefore, as a common wealth doth bosome enemies.

Suckers. A Sucker is a long, proud, and disorderly Cyon, growing straight vp (for pride of sap makes proud, long, and straight growth) cut of any lower parts of the tree, receiuing a great part of the sap, and bearing no fruit, till it haue tyrannized ouer the whole tree. These are like idle and great Drones amongst Bees; and proud and idle members in a common wealth.

Remedy. The remedy of this is, as of water-boughes, vnlesse he be growne greater then all the rest of the boughs, and then your Gardner (at your discretion) may leaue him for his boale, and take away all, or the most of the rest. If he be little, slip him, and set him, perhaps he will take: my fairest Apple-tree was such a Slip.

One principall top or bough, and remedy. One or two principall top boughes are as euill, in a manner, as Suckers, they rise of the same cause, and receiue the same remedy; yet these are more tolerable, because these beare fruit, yea the best: but Suckers of long doe not beare.

I know not how your tree should be faulty, if you reforme all your vices timely, and orderly. As these rules serue for dressing young trees and sets in the first planting: so may they well serue to helpe old trees, though not exactly to recouer them.

Instruments for dressing. The Instruments fittest for all these purposes, are most commonly: For the great trees an handsome long, light Ladder of Firpoles, a little, nimble, and strong armed Saw, and sharpe. For lesse Trees, a little and sharpe Hatchet, a broad mouthed Chesell, strong and sharpe, with an hand-beetle, your strong and sharpe Cleeuer, with a knock, & (which is a most necessary Instrument amongst little trees) a great hafted and sharpe Knife or Whittle. And as needfull is a Stoole on the top of a Ladder of eight or moe rungs, with two backe-feet, whereon you may safely and easefully stand to graffe, to dresse, and to gather fruit thus formed: The feet may be fast wedged in: but the Ladder must hang loose with two bands of iron. And thus much of dressing trees for fruit, formerly to profit.

Chap. 12.
Of Foyling.

Necessity of foiling. There is one thing yet very necessary for make your Orchard both better, and more lasting: Yea, so necessary, that without it your Orchard cannot last, nor prosper long, which is neglected generally both in precepts and in practice, viz. manuring with Foile: whereby it hapneth that when trees (amongst other euils) through want of fatnesse to feed them, become mossie, and in their growth are euill (or not) thriuing, it is either attributed to some wrong cause, as age (when indeed they are but young) or euill standing (stand they neuer so well) or such like, or else the cause is altogether vnknowne, and so not amended.

Trees great suckers. Can there be deuised any way by nature, or art, sooner or soundlier to seeke out, and take away the heart and strength of earth, then by great trees? Great bodies. Such great bodies cannot be sustained without great store of sap. What liuing body haue you greater then of trees? The great Sea monsters (whereof one came a land at *Teesmouth* in *Yorkeshire*, hard by vs, 18. yards in length, and neere as much in compasse) seeme hideous, huge, strange and monstrous, because they be indeed great: but especially because they are seldome seene: But a tree liuing, come to his growth and age, twice that length, and of a bulke neuer so great, besides his other parts, is not admired, because he is so commonly seene. And I doubt not, but if he were well regarded from his kirnell, by succeeding ages, to his full strength, the most of them would double their measure. About fifty yeeres agoe I heard by credible and constant report, That in *Brooham* Parke in *West moreland*, neere vnto *Penrith*, there lay a blowne Oake, whose trunke was so bigge, that two Horse men being the one on the one side, and the other on the other side, they could not one see another: to which if you adde his armes, boughs, and roots, and consider of his bignesse, what would he haue been, if preserued to the vantage. Also I read in the History of the *West-Indians*, out of *Peter Martyr*, that sixteene men taking hands one with another, were not able to fathome one of those trees about. Now Nature hauing giuen to such a faculty by large and infinite roots, taws and tangles, to draw immediately his sustenance from our common mother the Earth (which is like in this point to all other mothers that beare) hath also ordained that the tree ouer loden with fruit, and wanting sap to feed all she hath brought forth, will waine all she cannot feed, like a woman bringing forth moe children at once then she hath teats. See you not how trees especially, by kind being great, standing so thicke and close, that they cannot get plenty of sap, pine away all the grasse, weeds, lesser shrubs, and trees, yea and themselues also for want of vigor of sap? So that trees growing large, sucking the soile whereon they stand, continually, and amaine, and the foyzon of the earth that feeds them decaying (for what is there that wastes continually, that shall not haue end?) must either haue supply of sucker, or else leaue thriuing and growing. Some grounds will beare Corne while they be new, and no longer, because their crust is shallow, and not very good, and lying they scind and wash, and become barren. The ordinary Corne soiles continue not fertile, with fallowing and foyling, and the best requires supply, euen for the little body of Corne. How then can we thinke that any ground (how good soeuer) can containe bodies of such greatnesse, and such great feeding, without great plenty of Sap arising from good earth? This is one of the chiefe causes, why so many of our Orchards in *England* are so euill thriuing when they come to growth, and our fruit so bad. Men are loth to bestow much ground, and desire much fruit, and will neither set their trees in sufficient compasse, nor yet feed them with manure. Therefore of necessity Orchards must be foiled.

Time fit for foyling. The fittest time is, when your trees are growne great, and haue neere hand spread your earth, wanting new earth to sustaine them, which if they doe, they will seeke abroad for better earth, and shun that, which is barren (if they find better) as cattell euill pasturing. For nature hath taught euery creature to desire and seeke his owne good, and to auoid hurt. The best time of the yeere is at the Fall, that the Frost may bite and make it tender, and the Raine wash it to the roots. The Summer time is perillous if ye digge, because the sap fills amaine. Kind of foyle. The best kind of Foile is such as is fat, hot, and   tender. Your earth must be but lightly opened, that the dung may goe in, and wash away; and but shallow, lest you hurt the roots: and the spring closely and equally made plaine againe for feare of Suckers. I could wish, that after my trees haue fully possessed the soile of mine Orchard, that euery seuen yeeres at least, the soile were bespread with dung halfe a foot thicke at least. Puddle water out of the dunghill powred on plentifully, will not onely moisten but fatten especially in *Iune* and *Iuly*. If it be thicke and fat, and applied euery yeere, your Orchard shall need none other foiling. Your ground may lye so low at the Riuer side, that the floud standing some daies and nights thereon, shall saue you all this labour of foiling.

Chap. 13.
Of Annoyances.

A Chiefe helpe to make euery thing good, is to auoid the euils thereof: you shall neuer attaine to that good of your Orchard you looke for, vnlesse you haue a Gardner, that can discerne the diseases of your trees, and other annoyances of your Orchard, and find out the causes thereof, and know & apply fit remedies for the same. For be your ground, site, plants, and trees as you would wish, if they be wasted with hurtfull things, what haue you gained but your labour for your trauell? It is with an Orchard and euery tree, as with mans body, The best part of physicke for preseruation of health, is to foresee and cure diseases.

Two kinds of euils in an Orchard. All the diseases of an Orchard are of two sorts, either internall or externall. I call those inward hurts which breed on and in particular trees.

>1 Galles.
>2 Canker.
>3 Mosse.
>4 Weaknes in setting.
>5 Barke bound.
>6 Barke pild.
>7 Worme.
>8 Deadly wounds.

Galls. Galles, Canker, Mosse, weaknes, though they be diuers diseases: yet (howsoeuer Authors thinke otherwise) they rise all out of the same cause.

Galles we haue described with their cause and remedy, in the 11. Chapter vnder the name of fretters.

Canker. Canker is the consumption of any part of the tree, barke and wood, which also in the same place is deceiphered vnder the title of water-boughes.

Mosse. Mosse is sensibly seene and knowne of all, the cause is pointed out in the same Chapter, in the discourse of timber-wood, and partly also the remedy: but for Mosse adde this, that at any time in summer (the Spring is best) When the cause is remoued, with an Harecloth, immediatly after a showre of raine, rub off your Mosse, or with a peece of weed (if the Mosse abound) formed like a great knife.

Weaknesse in setting. Weaknesse in the setting of your fruit shall you finde there also in the same Chapter, and his remedy. All these flow from the want of roomth in good soile, wrong planting, Chap. 7. and euill or no dressing.

Barke-bound. Bark-bound (as I thinke) riseth of the same cause, and the best, & present remedy (the causes being taken away) is with your sharpe knife in the Spring, length-way to launch his bark throughout, on 3. or 4. sides of his boale.

Worme. The disease called the Worme is thus discernd: The barke will be hoald in diuers places like gall, the wood will die & dry, and you shall see easily the barke swell. It is verily to be thought, that therin is bred some worm I haue not yet thorowly sought it out, because I was neuer troubled therewithall: but onely haue seene such trees in diuers places. I thinke it a worme rather, because I see this disease in trees, bringing fruit of sweet taste, and the swelling shewes as much. Remedy. The remedy (as I coniecture) is so soone as you perceiue the wound, the next Spring cut it out barke and all, and apply Cowes pisse and vineger presently, and so twice or thrice a weeke for a moneths space: For I well perceiue, if you suffer it any time, it eates the tree or bough round, and so kils.

Since I first wrote this Treatise, I haue changed my mind concerning the disease called the worme, because I read in the History of the *West-Indians*, that their trees are not troubled with the disease called the worme or canker, which ariseth of a raw and euill concocted humor or sap, Witnesse *Pliny*, by reason their Country is more hot then ours, whereof I thinke the best remedy is (not disallowing the former, considering that the worme may breed by such an humor) warme standing, sound lopping and good dressing.

Barke pild. Bark-pild you shall find with his remedy in the 11. Chapter.

Wounds. Deadly wounds are when a mans Arborist wanting skill, cut off armes, boughes or branches an inch, or (as I see sometimes) an handfull, or halfe a foot or more from the body: These so cut cannot couer in any time with sap, and therefore they die, and dying they perish the heart, and so the tree becomes hollow, and with such a deadly wound cannot liue long.

Remedy. The remedy is, if you find him before he be perished, cut him close, as in the 11. Chapter: if he be hoald, cut him close, fill his wound, tho neuer so deepe, with morter well tempered & so close at the top his wound with a Seare-cloth doubled and nailed on, that no aire nor raine approach his wound. If he be not very old, and detaining, he will recouer, and the hole being closed, his wound within shall not hurt him for many yeeres.

Hurts on trees.
Ants, Earewigs, Caterpillars, and such like wormes. Hurts on your trees are chiefly Ants, Earewigs, and Caterpillars. Of Ants and Earewigs is said Chap. 10. Let there be no swarme of Pismires neere your tree-root, no not in your Orchard, turne them ouer in a frost, and powre in water, and you kill them.

For Caterpillars, the vigilant Fruterer shall soone espy their lodging by their web, or the decay of leaues eaten around about them. And being seene, they are easily destroyed with your hand, or rather (if your tree may spare it) take sprig and all: for the red peckled butterfly doth euer put them, being her sparm, among the tender spraies for better feeding, especially in drought, and tread them vnder your feet. I like nothing of smoke among my trees. Vnnaturall heates are nothing good for naturall trees. This for diseases of particular trees.

Externall euils. Externall hurts are either things naturall or artificiall. Naturall things, externally hurting Orchards.

	1 Deere.		1 Bulfinch.
	2 Goates.		2 Thrush.
	3 Sheepe.		3 Blackbird.
1 Beasts.	4 Hare.	2 Birds.	4 Crow.
	5 Cony.		5 Pye.
	6 Cattell.		
	7 Horse.		&c.

The other things are,
    1 Winds.

2 Cold.
   3 Trees.
   4 Weeds.
   5 Wormes.
      6 Mowles.
   7 Filth.
   8 Poysonfull smoke.
Externall wilfull euils are these.
   1 Walls.
   2 Trenches.
   3 Other works noisome done in or neere your Orchard.
   4 Euill Neighbours.
   5 A carelesse Master.
   6 An vndiscreet, negligent or no keeper.

See you here an whole Army of mischeifes banded in troupes against the most fruitfull trees the earth beares? assailing your good labours. Good things haue most enemies.

Remedy. A skilfull Fructerer must put so his helping hand, and disband and put them to flight.

Deere, &c. For the first ranke of beasts, besides your out strong fence, you must haue a faire and swift Greyhound, a stone-bow, gun, and if need require, an Apple with an hooke for a Deere, and an Hare-pipe for an Hare.

Birds. Your Cherries and other Berris when they be ripe, will draw all the Black-birds, Thrushes, and Maw Pies to your Orchard. The Bul-finch is a deuourer of your Fruit in the bud, I haue had whole trees shald out with them in Winter-time.

Remedy. The best remedy here is a Stone bow, a Piece, especially if you haue a Musket or Spar-hawke in Winter to make the Black bird stoope into a bush or hedge.

Other trees. The Gardner must cleanse his soile of all other trees: but fruit-trees aforesaid Chapter 2 for which it is ordained, and I would especially name Oakes, Elmes, Ashes, and such other great wood, but that I doubt it should be taken as an admission of lesser trees: for I admit of nothing to grow in mine Orchard but fruit and flowers. If sap can hardly be good to feed our fruit-trees, why should we allow of any other, especially those, that will becom their Masters, & wrong them in their liuelyhood.

Winds. And although we admit without the fence of Wall-nuts in most plaine places, Trees middle-most, and ashes or Okes, or Elmes vtmost, set in comely rowes equally distant with faire Allies twixt row and row to auoide the boisterous blasts of winds, and within them also others for Bees; yet wee admit none of these into your Orchard-plat: other remedy then this haue wee none against the nipping frosts. Frosts.

Weeds. Weeds in a fertile soile (because the generall curse is so) till your Trees grow great, will be noysome, and deforme your allies, walkes, beds, and squares, your vnder Gardners must labour to keepe all cleanly & handsome from them and all other filth with a Spade, weeding kniues, rake with iron teeth: a skrapple of Iron thus formed.

For Nettles and ground-Iuy after a showre.

Remedy. When weeds, straw, stickes and all other scrapings are gathered together, burne them not, but bury them vnder your crust in any place of your Orchard, and they will dye and fatten your ground.

Wormes.
Moales. Wormes and Moales open the earth, and let in aire to the roots of your trees, and deforme your squares and walkes, and feeding in the earth, being in number infinite, draw on barrennesse.

Remedy. Worms may be easily destroyed. Any Summer euening when it is darke, after a showre with a candle, you may fill bushels, but you must tred nimbly & where you cannot come to catch them so; sift the earth with coale ashes an inch or two thicknes, and that is a plague to them, so is sharpe grauell.

Moales will anger you, if your Gardner or some skilful Moale-catcher ease you not, especially hauing made their fortresses among the roots of your trees: you must watch her wel with a Moal spare, at morne, noon, and night, when you see her vtmost hill, cast a Trench betwixt her and her home (for she hath a principall mansion to dwell and breed in about *Aprill*, which you may discerne by a principall hill, wherein you may catch her, if you trench it round and sure, and watch well) or wheresoeuer you can discerne a single passage (for such she hath) there trench, and watch, and haue her.

Wilfull annoyances. Wilfull annoyances must be preuented and auoided by the loue of the Master and Fruterer, which they beare to their Orchard.

Remedy. Iustice and liberality will put away euill neighbours or euill neighbour-hood. And then if (God blesse and giue successe to your labours) I see not what hurt your Orchard can sustaine.

Chap. 14.
Of the age of Trees.

It is to be considered: All this Treatise of trees tends to this end, that men may loue and plant Orchards, whereunto there cannot be a better inducement then that they know (or at least be perswaded) that all that benefit they shall reape thereby, whether of pleasure or profit, shall not be for a day or a moneth, or one, or many (but many hundreth) yeeres. Of good things the greatest, and most durable is alwaies the best. The age of trees. If therefore out of reason grounded vpon experience, it be made (I thinke) manifest, but I am sure probable, that a fruit tree in such a soile and site, as is described so planted and trimmed and kept, as is afore appointed and duely foiled, shall dure 1000 yeeres, why should we not take paines, and be at two or three yeeres charges (for vnder seuen yeeres will an Orchard be perfected for the first planting, and in that time be brought to fruit) to reape such a commodity and so long lasting.

Gathered by reason out of experience. Let no man thinke this to be strange, but peruse and consider the reason. I haue Apple trees standing in my little Orchard, which I haue knowne these forty yeeres, whose age before my time I cannot learne, it is beyond memory, tho I haue enquired of diuers aged men of 80. yeeres and vpwards: these trees although come into my possession very euill ordered, mishapen, and one of them wounded to his heart, and that deadly (for I know it will be his death) with a wound, wherein I might haue put my foot in the heart of his bulke (now it is lesse) notwithstanding, with that small regard they haue had since, they so like, that I assure my selfe they are not come to their growth by more then 2. parts of 3. which I discerne not onely by their owne growth, but also by comparing them with the bulke of other trees. And I find them short (at least) by so many parts in bignesse, although I know those other fruit-trees to haue beene much hindred in their stature by euill guiding. Herehence I gather thus.

Parts of a trees age. If my trees be a hundred yeeres old, and yet want two hundred of their growth before they leaue encreasing, which make three hundred, then we must needs resolue, that this three hundred yeere are but the third part of a Trees life, because (as all things liuing besides)   so trees must haue allowed them for their increase one third, another third for their stand, and a third part of time also for their decay. All which time of a Tree amounts to nine hundred yeeres, three hundred for increase, three hundred for his stand, whereof we haue the terme stature, and three hundred for his decay, and yet I thinke (for we must coniecture by comparing, because no one man liueth to see the full age of trees) I am within the compasse of his age, supposing alwaies the foresaid meanes of preseruing his life. Consider the age of other liuing creatures. The Horse and moiled Oxe wrought to an vntimely death, yet double the time of their increase. A Dog likewise increaseth three, stanns three at least, end in as many (or rather moe) decayes.

Mans age. Euery liuing thing bestowes the least part of his age in his growth, and so must it needs be with trees. A man comes not to his full growth and strength (by common estimation) before thirty yeeres, and some slender and cleane bodies, not till forty, so long also stands his strength, & so long also must he haue allowed by course of nature to decay. Euer supposing that he be well kept with necessaries, and from and without straines, bruises, and all other dominyring diseases. I will not say vpon true report, that Physicke holds it possible, that a cleane body kept by these 3. Doctors, *Doctor Dyet*, *Doctor Quiet*, and *Doctor Merriman*, may liue neere a hundred yeeres. Neither will I here vrge the long yeeres of *Methushalah*, and those men of that time, because you will say, Mans dayes are shortned since the floud. But what hath shortned them? God for mans sinnes: but by meanes, as want of knowledge, euill gouernment, ryot, gluttony, drunkenesse, and (to be short) the encrease of the curse, our sinnes increasing in an iron and wicked age.

Now if a man, whose body is nothing (in a manner) but tender rottennesse, whose course of life cannot by any meanes, by counsell, restraint of Lawes, or punishment, nor hope of praise, profet, or eturnall glory, be kept within any bounds, who is degenerate cleane from his naturall feeding, to effeminate nicenesse, and cloying his body with excesse of meate, drinke, sleepe &c. and to whom nothing is so pleasant and so much desired as the causes of his owne death, as idlenesse, lust, &c. may liue to that age: I see not but a tree of a solide substance, not damnified by heate or cold, capable of, and subiect to any kinde of ordering or dressing that a man shall apply vnto him, feeding naturally, as from the beginning disburdened of all superfluities, eased of, and of his owne accord auoiding the causes that may annoy him, should double the life of a man, more then twice told; and yet naturall phylosophy, and the vniuersall consent of all Histories tell vs, that many other liuing creatures farre exceed man in the length of yeeres: As the Hart and the Rauen. Thus reporteth that famous *Roterodam* out of *Hesiodus*, and many other Historiographers. The testimony of *Cicero* in his booke *De Senectute*, is weighty to this purpose: that we must *in posteras ætates ferere arbores*, which can haue none other fence: but that our fruit-trees whereof he speakes, can endure for many ages.

What else are trees in comparison with the earth: but as haires to the body of a man? And it is certaine, without poisoning, euill and distemperate dyet, and vsage, or other such forcible cause, the haires dure with the body. That they be called excrements, it is by reason of their superfluous growth: (for cut them as often as you list, and they will still come to their naturall length) Not in respect of their substance, and nature. Haires endure long, and are an ornament and vse also to the body, as trees to the earth.

So that I resolue vpon good reason, that fruit-trees well ordered, may liue and like a thousand yeeres, and beare fruit, and the longer, the more, the greater, and the better, because his vigour is proud and stronger, when his yeeres are many: You shall see old trees put their buds and blossomes both sooner and more plentifully then young trees by much. And I sensibly perceiue my young trees to inlarge their fruit, as they grow greater, both for number and greatnesse. Young Heifers bring not forth the Calues so faire, neither are they so plentifull to milke, as when they become to be old Kine. No good Houswife will breed of a young but of an old bird-mother: It is so in all things naturally, therefore in trees.

The age of timber trees. And if fruit-trees last to this age, how many ages is it to be supposed, strong and huge timber-trees will last? whose huge bodies require the yeeres of diuers *Methushalaes*, before they end their dayes, whose sap is strong and bitter, whose barke is hard and thicke, and their substance solid and stiffe: all which are defences of health and long life. Their strength withstands all forcible winds, their sap of that quality is not subiect to wormes and tainting. Their barke receiues seldome or neuer by casualty any wound. And not onely so, but he is free from remoualls, which are the death of millions of trees, where as the fruit-tree in comparison is little, and often blowne downe, his sap sweet, easily and soone tainted, his barke tender, and soone wounded, and himselfe vsed by man, as man vseth himselfe, that is either   vnskilfully or carelessely.

Age of trees discerned. It is good for some purposes to regard the age of your fruit trees, which you may easily know, till they come to accomplish twenty yeeres, by his knots: Reckon from his root vp an arme and so to hys top-twig, and euery yeeres growth is distinguished from other by a knot, except lopping or remouing doe hinder.

Chap. 15.
Of gathering and keeping Fruit.

Generall Rule. Although it be an easie matter, when God shall send it, to gather and keepe fruit, yet are they certaine things worthy your regard. You must gather your fruit when it is ripe, and not before, else will it wither and be tough and sowre. All fruit generally are ripe, when they beginne to fall. For Trees doe as all other bearers doe, when their yong ones are ripe, they will waine them.

The Doue her Pigeons, the Cony her Rabbets, and women their children. Some fruit tree sometimes getting a taint in the setting with a frost or euill wind, will cast his fruit vntimely, but not before he leaue giuing them sap, or they leaue growing. Cherries, &c. Except from this foresaid rule, Cherries, Damsons and Bullies. The Cherry is ripe when he is sweld wholy red, and sweet: Damsons and Bulies not before the first frost.

Apples. Apples are knowne to be ripe, partly by their colour, growing towards a yellow, except the Leather-coat and some Peares and Greening.

When. Timely Summer fruit will be ready, some at Midsummer, most at Lammus for present vse; but generally noe keeping fruit before *Michal-tide*. Hard Winter fruit and Wardens longer.

Dry stalkes. Gather at the full of the Moone for keeping, gather dry for feare of rotting.

Gather the stalkes with all: for a little wound in fruit, is deadly: but not the stumpe, that must beare the next fruit, nor leaues, for moisture putrifies.

Seuerally. Gather euery kind seuerally by it selfe, for all will not keepe alike, and it is hard to discerne them, when they are mingled.

Ouerladen trees. If your trees be ouer-laden (as they will be, being ordered, as is before taught you) I like better of pulling some off (tho they be not ripe) neere the top end of the bough, then of propping by much, the rest shall be better fed. Propping puts the bough in danger, and frets it at least.

Instruments. Instruments: A long ladder of light Firre: A stoole-ladder as in the 11. Chapter. A gathering apron like a poake before you, made of purpose, or a Wallet hung on a bough, or a basket with a siue bottome, or skinne bottome, with Lathes or splinters vnder, hung in a rope to pull vp and downe: Bruises. bruise none, euery bruise is to fruit death: if you doe, vse them presently. An hooke to pull boughs to you is necessary, breake no boughes.

Keeping. For keeping, lay them in a dry Loft, the longest keeping Apples first and furthest on dry straw, on heapes ten or fourteene dayes, thicke, that they may sweat. Then dry them with a soft and cleane cloth, and lay them thinne abroad. Long keeping fruit would be turned once in a moneth softly: but not in nor immediately after frost. In a loft couer well with straw, but rather with chaffe or branne: For frost doth cause tender rottennesse.

Chap. 16.
Of Profits.

Now pause with your selfe, and view the end of all your labours in an Orchard: vnspeakable pleasure, and infinite commodity. The pleasure of an Orchard I referre to the last Chapter for the conclusion: and in this Chapter, a word or two of the profit, which thorowly to declare is past my skill: and I count it as if a man should attempt to adde light to the Sunne with a Candle, or number the Starres. No man that hath but a meane Orchard or iudgement but knowes, that the commodity of an Orchard is great: Neither would I speake of this, being a thing so manifest to all; but that I see, that through the carelesse lazinesse of men, it is a thing generally neglected. But let them know, that they lose hereby the chiefest good which belongs to house-keeping.

Compare the commodity that commeth of halfe an acre of ground, set with fruit-trees and hearbs, so as is prescribed, and an whole acre (say it be two) with Corne, or the best commodity you can wish, and the Orchard shall exceed by diuers degrees.

Cydar and Perry. In *France* and some other Countries, and in *England*, they make great vse of Cydar and Perry, thus made: Dresse euery Apple, the stalke, vpper end, and all galles away, stampe them, and straine them, and within 24. houres tun them vp into cleane, sweet, and sound vessels, for feare of euill ayre, which they will readily take: and if you hang a poakefull of Cloues, Mace, Nutmegs, Cinamon, Ginger, and pils of Lemmons in the midst of the vessell, it will make it as wholesome and pleasant as wine. The like vsage doth Perry require.

These drinks are very wholesome, they coole, purge, and preuent hot Agues. But I leaue this skill to Physicians.

Fruit. The benefit of your Fruit, Roots and Hearbs, though it were but to eate and sell, is much.

Waters. Waters distilled of Roses, Woodbind, Angelica, are both profitable and wondrous pleasant, and comfortable.

Conserue. Saffron and Licoras will yeeld you much Conserues and Preserues, are ornaments to your Feasts, health in your sicknesse, and a good helpe to your friend, and to your purse.

He that will not be moued with such vnspeakable profits, is well worthy to want, when others abound in plenty of good things.

Chap. 17.
Ornaments.

Me thinks hitherto we haue but a bare Orchard for fruit, and but halfe good, so long as it wants those comely Ornaments, that should giue beauty to all our labours, and make much for the honest delight of the owner and his friends.

Delight the chiefe end of Orchards. For it is not to be doubted: but as God hath giuen man things profitable, so hath he allowed him honest comfort, delight, and recreation in all the workes of his hands. Nay, all his labours vnder the Sunne without this are troubles, and vexation of mind: For what is greedy gaine, without delight, but moyling, and turmoyling in slauery? But comfortable delight, with content, is the good of euery thing, and the patterne of heauen. A morsell of bread with comfort, is better by much then a fat Oxe with vnquietnesse. An Orchard delightsome. And who can deny, but the principall end of an Orchard, is the honest delight of one wearied with the works of his lawfull calling? The very workes of, and in an Orchard and Garden, are better then the ease and rest of and from other labours. When God had made man after his owne Image, in a perfect state, and would haue him to represent himselfe in authority, tranquillity, and pleasure vpon the earth, he placed him in *Paradise*. An Orchard is Paradise. What was *Paradise*? but a Garden and Orchard of trees and hearbs, full of pleasure? and nothing there but delights. The gods of the earth, resembling the great God of heauen in authority, Maiestie, and abundance of all things, wherein is their most delight? Causes of wearisomnesse. and whither doe they withdraw themselues from the troublesome affaires of their estate, being tyred with the hearing and iudging of litigious Controuersies? choked (as it were) with the close ayres of their sumptuous buildings, their stomacks cloyed with variety of Banquets, their eares filled and ouerburthened with tedious discoursings? whither? but into their Orchards? Orchard is the remedy. made and prepared, dressed and destinated for that purpose, to renue and refresh their sences, and to call home their ouer-wearied spirits. Nay, it is (no doubt) a comfort to them, to set open their Cazements into a most delicate Garden and Orchard, whereby they may not onely see that, wherein they are so much delighted, but also to giue fresh, sweet, and pleasant ayre to their Galleries and Chambers.

All delight in Orchards. And looke, what these men do by reason of their greatnes and ability, prouoked with delight, the same doubtlesse would euery of vs doe, if power were answerable to our desires, whereby we shew manifestly, that of all other delights on earth, they that are taken by Orchards, are most excellent, and most agreeing with nature.

This delights all the senses. For whereas euery other pleasure commonly filles some one of our senses, and that onely, with delight, this makes all our sences swimme in pleasure, and that with infinite variety, ioyned with no lesse commodity.

Delighteth old age. That famous *Philosopher*, and matchlesse Orator, *M.T.C.* prescribeth nothing more fit, to take away the tediousnesse and heauy load of three or foure score yeeres, then the pleasure of an Orchard.

Causes of delight in an Orchard. What can your eye desire to see, your eares to hear, your mouth to tast, or your nose to smell, that is not to be had in an Orchard, with abundance and variety? What more delightsome then an infinite variety of sweet smelling flowers? decking with sundry colours, the greene mantle of the Earth, the vniuersall Mother of vs all, so by them bespotted, so dyed, that all the world cannot sample them, and wherein it is more fit to admire the Dyer, then imitate his workemanship. Colouring not onely the earth, but decking the ayre, and sweetning euery breath and spirit.

Flowers. The Rose red, damaske, veluet, and double double prouince Rose, the sweet muske Rose double and single, the double and single white Rose. The faire and sweet senting Woodbinde, double and single, and double double. Purple Cowslips, and double Cowslips, and double double Cowslips. Primerose double and single. The Violet nothing behinde the best, for smelling sweetly. A thousand more will prouoke your content.

Borders and squares. And all these, by the skill of your Gardner, so comely, and orderly placed in your Borders and Squares, and so intermingled, that none looking thereon, cannot but wonder, to see, what Nature corrected by Art can doe.

Mounts.
Whence you may shoote a Bucke.
Dyall.
Musique. When you behold in diuers corners of your Orchard *Mounts* of stone, or wood curiously wrought within and without, or of earth couered with fruit-trees: Kentish Cherry, Damsons, Plummes, &c. with staires of precious workmanship. And in some corner (or moe) a true Dyall or Clocke and some Anticke-workes and especially siluer-sounding Musique, mixt Instruments and voices, gracing all the rest: How will you be rapt with delight?
Walkes.
Seates. Large Walkes, broad and long, close and open, like the *Tempe* groues in *Thessalie*, raised with grauell and sand, hauing seats and bankes of Cammomile, all this delights the minde, and brings health to the body.

Order of trees. View now with delight the workes of your owne hands, your fruit-trees of all sorts, loaden with sweet blossomes, and fruit of all tasts, operations, and colours: your trees standing in comely order which way soeuer you looke.

Your borders on euery side hanging and drooping with Feberries, Raspberries, Barberries, Currens, and the rootes of your trees powdred with Strawberries, red, white, and greene, what a pleasure is this? Shape of men and beasts. Your Gardner can frame your lesser wood to the shape of men armed in the field, ready to giue battell: or swift running Greyhounds: or of well sented and true running

Hounds, to chase the Deere, or hunt the Hare. This kind of hunting shall not waste your corne, nor much your coyne.

Mazes. Mazes well framed a mans height, may perhaps make your friend wander in gathering of berries, till he cannot recouer himselfe without your helpe.

Bowle-Alley.
Buts. To haue occasion to exercise within your Orchard: it shall be a pleasure to haue a Bowling Alley, or rather   (which is more manly, and more healthfull) a paire of Buts, to stretch your armes.

Hearbes. Rosemary and sweete Eglantine are seemely ornaments about a Doore or Window, and so is Woodbinde.

Conduit. Looke Chapter 5, and you shall see the forme of a Conduite. If there were two or more, it were not amisse.

Riuer. And in mine opinion, I could highly commend your Orchard, if either through it, or hard by it there should runne a pleasant Riuer with siluer streames; you might sit in your Mount, and angle a peckled Trout, or fleightie Eele, or some other dainty Fish.
Moats. Or moats, whereon you might row with a Boate, and fish with Nettes.

Bees. Store of Bees in a dry and warme Bee-house, comely made of Fir-boords, to sing, and sit, and feede vpon your flowers and sprouts, make a pleasant noyse and sight. For cleanely and innocent Bees, of all other things, loue and become, and thriue in an Orchard. If they thriue (as they must needes, if your Gardiner bee skilfull, and loue them: for they loue their friends, and hate none but their enemies) they will, besides the pleasure, yeeld great profit, to pay him his wages Yea, the increase of twenty Stockes or Stooles, with other fees will keepe your Orchard.

You need not doubt their stings, for they hurt not whom they know, and they know their keeper and acquaintance. If you like not to come amongst them, you need not doubt them: for but neere their store, and in their owne defence, they will not fight, and in that case onely (and who can blame them?) they are manly, and fight desperately. Some (as that Honorable Lady   at *Hacknes*, whose name doth much grace mine Orchard) vse to make seates for them in the stone wall of their Orchard, or Garden, which is good, but wood is better.

Vine. A Vine ouer-shadowing a seate, is very comely, though her Grapes with vs ripe slowly.

Birds.
Nightingale. One chiefe grace that adornes an Orchard, I cannot let slip: A brood of Nightingales, who with their seuerall notes and tunes, with a strong delightsome voyce, out of a weake body, will beare you company night and day. She loues (and liues in) hots of woods in her hart. She will helpe you to cleanse your trees of Caterpillers, and all noysome wormes and flyes. Robin-red-brest.
Wren. The gentle Robin-red-brest will helpe her, and in winter in the coldest stormes will keepe a part. Neither will the silly Wren be behind in Summer, with her distinct whistle (like a sweete Recorder) to cheere your spirits.
Black-bird.
Thrush. The Black-bird and Threstle (for I take it the Thrush sings not, but deuoures) sing loudly in a *May* morning and delights the eare much (and you neede not want their company, if you haue ripe Cherries or Berries, and would as gladly as the rest do you pleasure:) But I had rather want their company than my fruit.

What shall I say? A thousand of pleasant delightes are attendant in an Orchard: and sooner shall I be weary, then I can recken the least part of that pleasure, which one that hath and loues an Orchard, may find therein.

What is there of all these few that I haue reckoned, which doth not please the eye, the eare, the smell, and taste? And by these sences as Organes, Pipes, and windowes, these delights are carried to refresh the gentle, generous, and noble mind.

Your owne labour. To conclude, what ioy may you haue, that you liuing to such an age, shall see the blessings of God on your labours while you liue, and leaue behind you to heires or successors (for God will make heires) such a worke, that many ages after your death, shall record your loue to their Countrey? And the rather, when you consider (*Chap. 14.*) to what length of time your worke is like to last.

FINIS.

THE COVNTRY HOVSE-VVIFES GARDEN.

Containing Rules for Hearbs and Seedes
of common vse, with their times and seasons,
when to set and sow them.
TOGETHER,
With the Husbandry of Bees, published with secrets
very necessary for euery House-wife.

As also diuerse new Knots for Gardens.

The Contents see at large in the last Page.

Genes. 2. 29.
I haue giuen vnto you euery Herbe, and euery tree, that shall be to you for meate.

LONDON,
Printed by *Nicholas Okes* for Iohn Harison, at the
golden Vnicorne in Pater-noster-row. 1631.

THE COVNTRY HOVSVVIFES GARDEN.

Chap. 1.
The Soyle.

The soyle of an Orchard and Garden, differ onely in these three points: Dry. First, the Gardens soyle would be somewhat dryer, because hearbes being more tender then trees, can neither abide moisture nor drought, in such excessiue measure, as trees; and therefore hauing a dryer soyle, the remedy is easie against drought, if need be: water soundly, which may be done with small labour, the compasse of a Garden being nothing so great, as of an Orchard, and this is the cause (if they know it) that Gardners raise their squares:Hops. but if moysture trouble you, I see no remedy without a generall danger, except in Hops, which delight much in a low and sappy earth.

Plaine. Secondly, the soyle of a Garden would be plaine and leuell, at least euery square (for we purpose the square to be the fittest forme) the reason: the earth of a garden wanting such helpes, as should stay the water, which an orchard hath, and the rootes of hearbes being short, and not able to fetch their liquor from the bottome, are more annoyed by drought, and the soyle being mellow and loose, is soone either washt away, or sends out his heart by too much drenching and washing.

Thirdly, if a garden soyle be not cleere of weedes, and namely, of grasse, the hearbes shall neuer thriue: for how should good hearbes prosper, when euill weeds waxe so fast: considering good hearbes are tender in respect of euill weedes: these being strengthened by nature, and the other by art? Gardens haue small place in comparison, and therefore may be more easily be fallowed, at the least one halfe yeare before, and the better dressed after it is framed. And you shall finde that cleane keeping doth not onely auoide danger of gathering weedes, but also is a speciall ornament, and leaues more plentifull sap for your tender hearbes.

Chap. 2.
Of the Sites.

I cannot see in any sort, how the site of the one should not be good, and fit for the other: The ends of both being one, good, wholesome, and much fruit ioyned with delight, vnlesse trees be more able to abide the nipping frostes than tender hearbes: but I am sure, the flowers of trees are as soone perished with cold, as any hearbe except Pumpions, and Melons.

Chap. 3.
Of the Forme.

Let that which is sayd in the Orchards forme, suffice for a garden in generall: but for speciall formes in squares, they are as many, as there are diuices in Gardners braines. Neither is the wit and art of a skilfull Gardner in this poynt not to be commended, that can worke more variety for breeding of more delightsome choyce, and of all those things, where the owner is able and desirous to be satisfied. The number of formes, Mazes and Knots is so great, and men are so diuersly delighted, that I leaue euery House-wife to her selfe, especially seeing to set downe many, had bene but to fill much paper; yet lest I depriue her of all delight and direction, let her view these few, choyse, new formes, and note this generally, that all plots are square, and all are bordered about with Priuit, Raisins, Fea-berries, Roses, Thorne, Rosemary, Bee-flowers, Isop, Sage, or such like.

The ground plot for Knots.

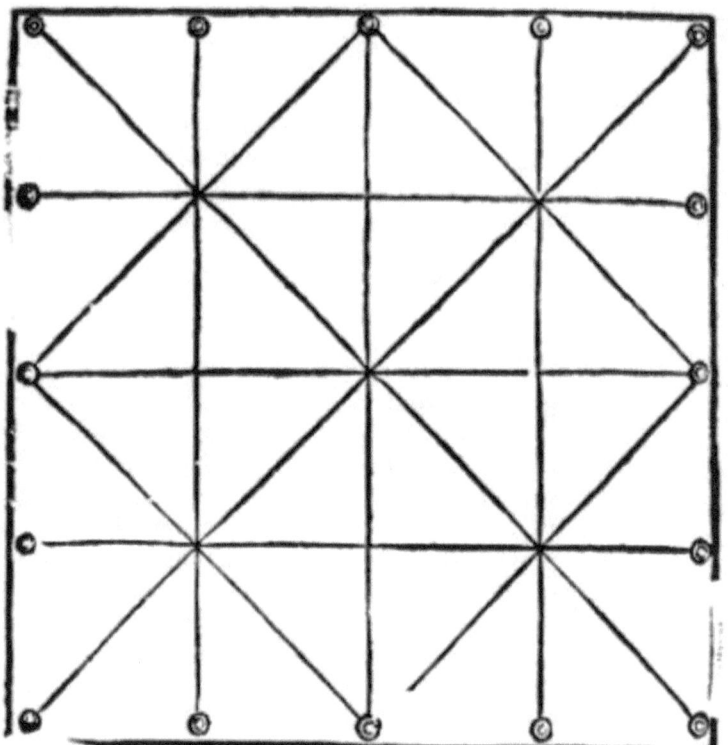

Cinkfoyle.

Flower-deluce.

The Tre-oyle.

The Fret.

Lozenges.

Crosse-bow.

Diamond.

Ouall.

Maze.

Chap. 4.
Of the Quantity.

A Garden requireth not so large a scope of ground as an Orchard, both in regard of the much weeding, dressing and remouing, and also the paines in a Garden is not so well repaied home, as in an Orchard. It is to be graunted, that the Kitchin garden doth yeeld rich gaines by berries, roots, cabbages, &c. yet these are no way comparable to the fruits of a rich Orchard: but notwithstanding I am of opinion, that it were better for *England*, that we had more Orchards and Gardens, and more large. And therefore we leaue the quantity to euery mans ability and will.

Chap. 5.
Of Fence.

Seeing we allow Gardens in Orchard plots, and the benefit of a Garden is much, they both require a strong and shrowding fence. Therefore leauing this, let vs come to the hearbes themselues, which must be the fruit of all these labours.

Chap. 6.
Of two Gardens.

Hearbes are of two sorts, and therefore it is meete (they requiring diuers manners of Husbandry) that we haue two Gardens: A garden for flowers, and a Kitchen garden: or a Summer garden: not that we meane so perfect a distinction, that the Garden for flowers should or can be without hearbes good for the Kitchen, or the Kitchen garden should want flowers, nor on the contrary: but for the most part they would be seuered: first, because your Garden flowers shall suffer some disgrace, if among them you intermingle Onions, Parsnips, &c. Secondly, your Garden that is durable, must be of one forme: but that, which is for your Kitchens vse, must yeeld daily rootes, or other hearbes, and suffer deformity. Thirdly, the hearbs of both will not be both alike ready, at one time, either for gathering, or re-mouing. First therefore

Of the Summer Garden.

These hearbs and flowers are comely and durable for squares and knots and all to set at *Michael-*tide*, or somewhat before, that they may be setled in, and taken with the ground before winter, though they may be set, especially sowne in the spring.

Roses of all sorts (spoken of in the Orchard) must be set. Some vie to set slips and twine them, which sometimes, but seldome thriue all.

Rosemary, Lauender, Bee-flowers, Isop, Sage, Time, Cowslips, Pyony, Dasies, Cloue Gilliflowers, Pinkes, Sothernwood, Lillies, of all which hereafter.

Of the Kitchen Garden.

Though your Garden for flowers doth in a sort peculiarly challenge to it seise a profit, and exquisite forme to the eyes, yet you may not altogether neglect this, where your hearbes for the pot do growe. And therefore, some here make comely borders with the hearbes aforesayd. The rather because aboundance of Roses and Lauender yeeld much profit, and comfort to the sences: Rose-water and Lauender, the one cordial (as also the Violets, Burrage, and Buglas) the other reuiuing the spirits by the sence of smelling: both most durable for smell, both in flowers and water: you need not here raise your beds, as in the other garden, because Summer towards, will not let too much wet annoy you.

And these hearbes require more moysture: yet must you haue your beds diuided, that you may goe betwixt to weede, and somewhat forme would be expected: To which it auaileth, that you place your herbes of biggest growth, by walles, or in borders, as Fenell, &c. and the lowest in the middest, as Saffron, Strawberries, Onions, &c.

Chap. 7.
Diuision of hearbs.

Garden hearbs are innumerable, yet these are common and sufficient for our country House-wifes.

Hearbs of greatest growth.

Fenell, Anglica, Tansie, Hollihock, Louage, Elly Campane, French mallows, Lillies, French poppy, Endiue, Succory and Clary.

Herbes of middle growth.

Burrage, Buglas, Parsley, sweet Sicilly, Floure-de-luce, Stocke Gilliflowers, Wall-flowers, Anniseedes, Coriander, Feather fewell, Marigolds, Oculus Christi, Langdibeefe, Alexanders, Carduus Benedictus.

Hearbes of smallest growth.

Pansy, or Harts-ease, Coast Margeram, Sauery, Strawberries, Saffron, Lycoras, Daffadowndillies, Leekes, Chiues, Chibals, Skerots, Onions, Batchellors buttons, Dasies, Peniroyall.

Hitherto I haue onely reckoned vp, and put in this ranke, some hearbs. Their Husbandry follow each in an Alphabeticall order, the better to be found.

Chap. 8.
Husbandry of Herbes.

A lexanders are to be renewed as *Angelica*. It is a timely Pot-hearbe.

Anglica is renued with his seede, whereof he beareth plenty the second yeare, and so dieth. You may remoue the rootes the first yeare. The leaues distilled, yeeld water soueraigne to expell paine from the stomacke. The roote dried taken in the fall, stoppeth the poares against infections.

Annyseedes make their growth, and beareth seeds the first yeere, and dieth as *Coriander*: it is good for opening the pipes, and it is vsed in Comfits.

Artichoakes are renewed by diuiding the rootes into sets, in *March*, euery third or fourth yeare. They require a seuerall vsage, and therefore a seuerall whole plot by themselues, especially considering they are plentifull of fruite much desired.

Burrage and *Buglas*, two Cordials, renue themselues by seed yearely, which is hard to be gathered: they are exceeding good Pot-hearbes, good for Bees, and most comfortable for the heart and stomacke, as Quinces and Wardens.

Camomile, set rootes in bankes and walkes. It is sweete smelling, qualifying head-ach.

Cabbages require great roome, they seed the second yeare: sow them in *February*, remoue them when the plants are an handfull long, set deepe and wet. Looke well in drought for the white Caterpillers worme, the spaunes vnder the leafe closely; for euery liuing Creature doth seeke foode and quiet shelter, and growing quicke, they draw to, and eate the heart: you may finde them in a rainy deawy morning.

It is a good Pothearbe, and of this hearbe called *Cole* our Countrie House-wiues giue their pottage their name, and call them *Caell*.

Carduus Benedictus, or blessed thistle, seeds and dyes the first yeere, the excellent vertue thereof I referre to Herbals, for we are Gardiners, not Physitians.

Carrets are sowne late in *Aprill* or *May*, as Turneps, else they seede the first yeere, and then their roots are naught: the second yeere they dye, their roots grow great, and require large roome.

Chibals or *Chiues* haue their roots parted, as Garlick, Lillies, &c. and so are they set euery third or fourth yeere: a good pot-hearb opening, but euill for the eies.

Clarie is sowne, it seeds the second yeere, and dyes. It is somewhat harsh in taste, a little in pottage is good, it strengtheneth the reines.

Coast, Roote parted make sets in *March*: it beares the second yeere: it is vsed in Ale in *May*.

Coriander is for vsage and vses, much like Anniseeds.

Daffadowndillies haue their roots parted, and set once in three or foure yeere, or longer time. They flower timely, and after *Midsummer*, are scarcely seene. They are more for ornament, then for vse, so are Daisies.

Daisie-rootes parted and set, as Flowre-deluce and Camomile, when you see them grow too thicke or decay. They be good to keepe vp, and strengthen the edges of your borders, as Pinkes, they be red, white, mixt.

Ellycampane root is long lasting, as is the Louage, it seeds yeerely, you may diuide the root, and set the roote, taken in VVinter it is good (being dryed, powdered and drunke) to kill itches.

Endiue and *Succory* are much like in nature, shape, and vse, they renue themselues by seed, as Fennell, and other hearbs. You may remoue them before they put forth shankes, a good Pot-hearbe.

Fennell is renued, either by the seeds (which it beareth the second yeere, and so yeerely in great abundance) sowne in the fall or Spring, or by diuiding one root into many Sets, as Artichoke, it is long of growth and life. You may remoue the roote vnshankt. It is exceeding good for the eyes, distilled, or any otherwise taken: it is vsed in dressing Hiues for swarmes, a very good Pot-hearbe, or for Sallets.

Fetherfewle shakes seed. Good against a shaking Feuer, taken in a posset drinke fasting.

Flower-deluce, long lasting. Diuide his roots, and set: the roots dryed haue a sweet smell.

Garlicke may be set an handfull distance, two inches deepe, in the edge of your beds. Part the heads into seuerall cloues, and euery cloue set in the latter end of *February*, will increase to a great head before *September*: good for opening, euill for eyes: when the blade is long, fast two & two together, the heads will be bigger.

Hollyhocke riseth high, seedeth and dyeth: the chiefe vse I know is ornament.

Isop is reasonable long lasting: young roots are good set, slips better. A good pot-hearbe.

Iuly-flowers, commonly called *Gilly-flowers*, or *Cloue-Iuly-flowers* (I call them so, because they flowre in *Iuly*) they haue the name of *Cloues*, of their sent. I may well call them the King of flowers (except the Rose) and the best sort of them are called *Queene-Iuly flowers*. I haue of them nine or ten seuerall colours, and diuers of them as big as Roses; of all flowers (saue the Damaske Rose) they are the most pleasant to sight and smell: they last not past three or foure yeeres vnremoued. Take the slips (without shanks) and set any time, saue in extreme frost, but especially at *Michael tide*. Their vse is much in ornament, and comforting the spirits, by the sence of smelling.

Iuly flowers of the wall, or wall-*Iuly-flowers*, wall-flowers, or Bee-flowers, or Winter-*Iuly-flowers*, because growing in the walles, euen in Winter, and good for Bees, will grow euen in stone walls, they will seeme dead in Summer, and yet reuiue in Winter. They yeeld seed plentifully, which you may sow at any time, or in any broken earth, especially on the top of a mud-wall, but moist, you may set the root before it be brancht, euery slip that is not flowr'd will take root, or crop him in Summer, and he will flower in Winter: but his Winter-seed is vntimely. This and Palmes are exceeding good, and timely for Bees.

Leekes yeeld seed the second yeere, vnremoued and die, vnlesse you remoue them, vsuall to eate with salt and bread, as Onyons alwaies greene, good pot-hearb, euill for the eyes.

Lauendar spike would be remoued within 7 yeeres, or eight at the most. Slips twined as Isop and Sage, would take best at *Michael-tide*. This flower is good for Bees, most comfortable for smelling, except Roses; and kept dry, is as strong after a yeere, and when it is gathered. The water of this is comfortable.

White *Lauendar* would be remoued sooner.

Lettice yeelds seed the first yeere, and dyes: sow betime, and if you would haue them *Cabbage* for Sallets, remoue them as you doe *Cabbage*. They are vsuall in Sallets, and the pot.

Lillies white and red, remoued once in three or foure yeeres their roots yeeld many Sets, like the Garlicke, *Michael-tide* is the best: they grow high, after they get roote: these roots are good to breake a Byle, as are Mallowes and Sorrell.

Mallowes, French or gagged, the first or second yeere, seed plentifully: sow in *March*, or before, they are good for the house-wifes pot, or to breake a bunch.

Marigolds most commonly come of seed, you may remoue the Plants, when they be two inches long. The double Marigold, being as bigge as a little Rose, is good for shew. They are a good Pot-hearbe.

Oculus Christi, or Christs eye, seeds and dyes the first or second yeere: you may remoue the yong Plants, but seed is better: one of these seeds put into the eye, within three or foure houres will gather a thicke skinne, cleere the eye, and bolt it selfe forth without hurt to the eye. A good Pot-hearbe.

Onyons are sowne in *February*, they are gathered at *Michael-tide*, and all the Summer long, for Sallets; as also young Parsly, Sage,

Chibals, Lettice, sweet Sicily, Fennell, &c. good alone, or with meate as Mutton, &c. for sauce, especially for the pot.

Parsly sow the first yeere, and vse the next yeere: it seedes plentifully, an hearbe of much vse, as sweet Sicily is. The seed and roots are good against the Stone.

Parsneps require and whole plot, they be plentifull and common: sow them in *February*, the Kings (that is in the middle) seed broadest and reddest. Parsneps are sustenance for a strong stomacke, not good for euill eies: When they couer the earth in a drought, to tread the tops, make the rootes bigger.

Peny-royall, or Pudding Grasse, creepes along the ground like ground Iuie. It lasts long, like Daisies, because it puts and spreads dayly new roots. Diuide, and remoue the roots, it hath a pleasant taste and smell, good for the pot, or hackt meate, or Haggas Pudding.

Pumpions: Set seedes with your finger, a finger deepe, late in *March*, and so soone as they appeare, euery night if you doubt frost, couer them, and water them continually out of a water-pot: they be very tender,   their fruit is great and waterish.

French poppy beareth a faire flower, and the Seed will make you sleepe.

Raddish is sauce for cloyed stomacks, as Capers, Oliues, and Cucumbers, cast the seeds all Summer long here and there, and you shall haue them alwaies young and fresh.

Rosemary, the grace of hearbs here in *England*, in other Countries common. To set slips immediately after *Lammas*, is the surest way. Seede sowne may proue well, so they be sowne in hot weather, somewhat moist, and good earth: for the hearbe, though great, is nesh and tender (as I take it) brought from hot Countries to vs in the cold North: set thinne. It becomes a Window well. The vse is much in meates, more in Physicke, most for Bees.

Rue, or Hearbe of Grace, continually greene, the slips are set. It lasts long as Rosemary, Sothernwood, &c. too strong for mine Housewifes pot, vnlesse she will brue Ale therewith, against the Plague: let him not seede, if you will haue him last.

Saffron euery third yeere his roots would be remoued at *Midsummer*: for when all other hearbs grow most, it dyeth. It flowreth at *Michael-tide*, and groweth all Winter: keepe his flowers from birds in the morning, & gather the yellow (or they shape much like Lillies) dry, and after dry them: they be precious, expelling diseases from the heart and stomacke.

Sauery seeds and dyes the first yeere, good for my Housewifes pot and pye.

Sage: set slips in *May*, and they grow aye: Let it not seed it will last the longer. The vse is much and common. The Monkish Prouerbe is *tritum*:

Cur moritur homo, cum saluia crescit in horto?

Skerots, roots are set when they be parted, as *Pyonie*, and Flowerdeluce at *Michael-tide*: the roote is but small and very sweet. I know none other speciall vse but the Table.

Sweet *Sicily*, long lasting, pleasantly tasting, either the seed sowne, or the root parted, and remoued, makes increase, it is of like vse with Parsly.

Strawberries long lasting, set roots at *Michael-tide* or the Spring, they be red, white and greene, and ripe, when they be great and soft, some by *Midsummer* with vs. The vse is: they will coole my Housewife well, if they be put in Wine or Creame with Sugar.

Time, both seeds, slips and rootes are good. If it seed not, it will last three or foure yeeres or more, it smelleth comfortably. It hath much vse: namely, in all cold meats, it is good for Bees.

Turnep is sowne. In the second yeere they beare plenty of seed: they require the same time of sowing that Carrets doe: they are sicke of the same disease that Cabbages be. The roots increaseth much, it is most wholesome, if it be sowne in a good and well tempered earth: Soueraigne for eyes and Bees.

I reckon these hearbs onely, because I teach my Countrey Housewife, not skilfull Artists, and it should be an endlesse labour, and would make the matter tedious to reckon vp *Landtheefe, Stocke-Iuly-flowers, Charuall, Valerian, Go-to bed at noone, Piony, Licoras, Tansie, Garden mints, Germander, Centaurie,* and a thousand such physicke Hearbs. Let her first grow cunning in this, and then she may enlarge her Garden as her skill and ability increaseth. And to helpe her the more, I haue set her downe these obseruations.

Chap. 9.
Generall Rules in Gardening.

In the South parts Gardening may be more timely, and more safely done, then with vs in *Yorkeshire*, because our ayre is not so fauourable, nor our ground so good.

2 Secondly most seeds shakt, by turning the good earth, are renued, their mother the earth keeping them in her bowels, till the Sunne their Father can reach them with his heat.

3 In setting hearbs, leaue no top more then an handfull aboue the ground, nor more then a foot vnder the earth.

4 Twine the roots of those slips you set, if they will abide it. Gillyflowers are too tender.

5 Set moist, and sowe dry.

6 Set slips without shankes any time, except at *Midsummer*, and in frosts.

7 Seeding spoiles the most roots, as drawing the heart and sap from the root.

8 Gather for the pot and medicines, hearbs tender and greene, the sap being in the top, but in Winter the root is best.

9 All the hearbs in the Garden for flowers, would once in seuen yeeres be renued, or soundly watered with puddle water, except Rosemary.

10 In all your Gardens and Orchards, bankes and seates of Camomile, Peny-royall, Daisies and Violets, are seemely and comfortable.

11 These require whole plots: Artichokes, Cabbages, Turneps, Parsneps, Onyons, Carrets, and (if you will) Saffron and Scerrits.

12 Gather all your seeds, dead, ripe, and dry.

13 Lay no dung to the roots of your hearbs, as vsually they doe: for dung not melted is too hot, euen for trees.

14 Thin setting and sewing (so the rootes stand not past a foot distance) is profitable, for the hearbs will like the better. Greater hearbs would haue more distance.

15 Set and sow hearbs in their time of growth (except at *Midsummer*, for then they are too too tender) but trees in their time of rest.

16 A good Housewife may, and will gather store of hearbs for the pot, about *Lammas*, and dry them, and pownd them, and in Winter they will make good seruice.

Thus haue I lined out a Garden to our Countrey Housewiues, and giuen them rules for common hearbs. If any of them (as sometimes they are) be knotty, I referre them to Chap. 3. The skill and paines of weeding the Garden with weeding kniues or fingers, I refer to themselues, and their maides, willing them to take the opportunitie after a showre of raine: withall I aduise the Mistresse, either be present her selfe, or to teach her maides to know hearbs from weeds.

Chap. 10.
The Husbandry of Bees.

There remaineth one necessary thing to be prescribed, which in mine opinion makes as much for ornament as either Flowers, or forme, or cleanlinesse, and I am sure as commodious as any of, or all the rest: which is Bees, well ordered. And I will not account her any of my good House-wiues, that wanteth either Bees or skilfulnesse about them. And though I knowe some haue written well and truely, and others more plentifully vpon this theame: yet somewhat haue I learned by experience (being a Bee-maister my selfe) which hitherto I cannot finde put into writing, for which I thinke our House-wiues will count themselues beholding vnto me.

Bee-house. The first thing that a Gardiner about Bees must be carefull for, is an house not stakes and stones abroad, *Sub dio*: for stakes rot and reele, raine and weather eate your hiues, and couers, and cold most of all is hurtfull for your Bees. Therefore you must haue an house made along, a sure dry wall in your Garden, neere, or in your Orchard: for Bees loue flowers and wood with their hearts.

This is the forme, a Frame standing on posts with a Floore (if you would haue it hold more Hiues, two Floores boorded) layd on bearers, and backe posts, couered ouer with boords, slate-wise.

Let the floores be without holes or clifts, least in casting time, the Bees lye out, and loyter.

And though your Hiues stand within an hand breadth the one of another: yet will Bees know their home.

In this Frame may your Bees stand drye and warme, especially if you make doores like doores of windows to shroud them in winter, as in an house: prouided you leaue the hiues mouths open. I my self haue deuised such an house, and I find that it keeps and strengthens my Bees much, and my hiues will last six to one.

Hiues. M. *Markham* commends Hiues of wood. I discommend them not: but straw Hiues are in vse with vs, and I thinke with all the world, which I commend for nimblenesse, closenesse, warmnesse and drinesse. Bees loue no externall motions of dawbing or such like. Sometimes occasion shall be offered to lift and turne Hiues, as shall appeare hereafter. One light entire hiue of straw in that case is better, then one that is dawbed, weighty and cumbersome. I wish euery hiue, for a keeping swarme, to hold three pecks at least in measure. For too little Hiues procure Bees, in casting time, either to lye out, and loyter, or else to cast before they be ripe and strong, and so make weake swarmes and vntimely: Whereas if they haue roome sufficient, they ripen timely, and casting seasonably, are strong, and fit for labour presently. Neither would the hiue be too too great, for then they loyter, and waste meate and time.

Hiuing of Bees. Your Bees delight in wood, for feeding, especially for casting: therefore want not an Orchard. A *Mayes* swarme is worth a Mares Foale: if they want wood, they be in danger of flying away. Any time before *Midsummer* is good, for casting and timely before *Iuly* is not euill. I much like M *Markhams* opinion for hiuing a swarme in combes of a dead or forsaken hiue, so they be fresh & cleanly. To thinke that a swarme of your owne, or others, will of it selfe come into such an hiue, is a meere conceit. *Experto crede Roberto.* His smearing with honey, is to no purpose, for the other Bees will eate it vp. If your swarme knit in the top of a tree, as they will, if the winde beate them not to fall   downe: let the stoole or ladder described in the Orchard, doe you seruice.

Spelkes. The lesse your Spelkes are, the lesse is the waste of your honey, and the more easily will they draw, when you take your Bees. Foure Spelkes athwart, and one top Spelke are sufficient. The Bees will fasten their combes to the Hiue. A little honey is good: but if you want, Fennell will serue to rub your Hiue withall. The Hiue being drest and ready spelkt, rubd and the hole made for their passage (I vse no hole in the Hiue, but a piece of wood hoal'd to saue the hiue & keep out Mice) shake in your Bees, or the most of them (for all commonly you cannot get) the remainder will follow. Many vse smoke, Nettles, &c. which I vtterly dislike: for Bees loue not to be molested. Ringing in the time of casting is a meere fancie, violent handling of them is simply euill, because Bees of all other creatures, loue cleanlinesse and peace. Therefore handle them leasurely & quietly, and their Keeper whom they know, may do with them, what he will, without hurt: Being hiued at night, bring them to their seat. Set your hiues all of one yeere together.

Signes of breeding, if they be strong:

1 They will auoid dead young Bees and Droanes.

2 They will sweat in the morning, till it runne from them; alwaies when they be strong.

Signes of casting.

1 They will fly Droanes, by reason of heat.

2 The young swarme will once or twice in some faire season, come forth mustering, as though they would cast, to proue themselues, and goe in againe.

3 The night before they cast, if you lay your eare to the Hiues mouth, yo shall heare two or three, but especially one aboue the rest, cry, Vp, vp, vp; or, Tout, tout, tout, like a trumpet, sounding the alarum to the battell.

Much descanting there is, of, and about the Master-Bee, and their degrees, order and gouernment: but the truth in this point is rather imagined, then demonstrated. There are some coniectures of it, *viz.* we see in the combs diuers greater houses then the rest, & we heare commonly the night before they cast, sometimes one Bee, sometimes two, or more Bees, giue a lowd and seuerall sound from the rest, and sometimes Bees of greater bodies then the common sort: but what of all this? I leane not on coniectures, but loue to set downe that I know to be true, and leaue these things to them that loue to diuine. Keepe none weake, for it is hazard, oftentimes with losse: Feeding will not helpe them: for being weake, they cannot come downe to meate, or if they come downe, they dye, because Bees weake cannot abide cold. If none of these, yet will the other Bees being strong, smell the honey, and come and spoile, and kill them. Catching. Some helpe is in casting time, to put two weake swarmes together, or as M. *Markham* well saith: Let not them cast late, by raising them with wood or stone: but with impes (say I.) An impe is three or foure wreathes, wrought as the hiue, the same compasse, to rase the hiue withall: Clustering. but by experience in tryall, I haue found out a better way by Clustering, for late or weake swarmes hitherto not found out of any that I know. That is this: After casting time, if I haue any stocke proud, and hindered from timely casting, with former Winters pouerty, or euill weather in casting time, with two handles and crookes, fitted for the purpose, I turne vp that stocke so pestred with Bees, and set it on the crowne, vpon which so turned with the mouth vpward, I place another empty hiue well drest, and spelkt, into which without any labour, the Swarme that would not depart, and cast, will presently ascend, because the old Bees haue this qualitie (as all other breeding creatures haue) to expell the young, when they haue brought them vp.

There will the swarme build as kindely, as if they had of themselues beene cast. But bee sure you lay betwixt the Hiues some straight and cleanly sticke or stickes, or rather a boord with holes, to keepe them asunder: otherwise they will ioyne their workes together so fast, that they cannot be parted. If you so keepe them asunder at *Michael-tide*, if you like the weight of your swarme (for the goodnesse of swarmes is tryed by weight) so catched, you may set it by for a stocke to keepe. Take heed in any case the combes be not broken, for then the other Bees will smell the honey, and spoyle them. This haue I tryed to be very profitable for the sauing of Bees. The Instrument hath this forme. The great straight piece is wood, the rest are iron claspes and nailes, the claspes are loose in the Stapes: Two men with two of these fastened to the Hiue, will easily turne it vp.

They gather not till *Iuly*; for then they be discharged of their young, or else they are become now strong to labour, and now sap in flowers is strong and proud: by reason of time, and force of Sunne. And now also in the North (and not before) the hearbs of greatest vigour put their Flowers; As Beanes, Fennell, Burrage, Rape, &c.

The most sensible weather for them, is heat and drought, because the nesh Bee can neither abide cold or wet: and showres (which they well fore-see) doe interrupt their labours, vnlesse they fall on the night, and so they further them.

Droanes. After casting time, you shall benefit your stockes much, if you helpe them to kill their Droanes, which by all probability and iudgement, are an idle kind of Bees, and wastefull. Some say they breed and haue seene young Droanes in taking their honey, which I know is true. But I am of opinion, that there are also Bees which haue lost their stings, and so being, as it were gelded, become idle and great. There is great vse of them: *Deus, et natura nihil fecit frustra*. They hate the Bees, and cause them cast the sooner. They neuer come foorth but when they be ouer heated. They neuer come home loaden. After casting time, and when the Bees want meate, you shall see the labouring Bees fasten on them, two, three, or foure at once, as if they were theeues to be led to the gallowes, and killing them, they cast out, and draw them farre from home, as hatefull enemies. Our Housewife, if she be the Keeper of her owne Bees (as she had need to be) may with her bare hand in the heate of the day, safely destroy them in the hiues mouth. Some vse towards night, in a hot day, to set before the mouth of the hiue a thin board, with little holes, in at which the lesser Bees may enter, but not the Droanes, so that you may kill them at your pleasure.

Annoyances. Snayles spoile them by night like theeues: they come so quietly, and are so fast, that the Bees feare them not. Looke earely and late, especially in a rainie or dewey euening or morning.

Mice are no lesse hurtfull, and the rather to hiues of straw: and therefore couerings of straw draw them. They will in either at the mouth, or sheere themselues an hole. The remedy is good Cats, Rats-bane and watching.

The cleanly Bee hateth the smoake as poison, therefore let your Bees stand neerer your garden then your Brew-house or Kitchen.

They say Sparrowes and Swallowes are enemies to Bees, but I see it not.

More hiues perish by Winters cold, then by all other hurts: for the Bee is tender and nice, and onely liues in warme weather, and dyes in cold: And therefore let my Housewife be perswaded, that a warme dry house before described, is the chiefest helpe she can make her Bees against this, and many more mischiefes. Many vse against cold in Winter, to stop vp their hiue close, and some set them in houses, perswading themselues, that thereby they relieue their Bees. First, tossing and mouing is hurtfull. Secondly, in houses, going, knocking, and shaking is noysome. Thirdly, too much heate in an house is vnnaturall for them: but lastly, and especially, Bees cannot abide to be stopt close vp. For at euery warme season of the Sunne they reuiue, and liuing eate, and eating must needs purge abroad, (in her house) the cleanly Bee will not purge her selfe. Iudge you what it is for any liuing creature, not to disburden nature. Being shut vp in calme seasons, lay your care to the Hiue, and you shall heare them yarme and yell, as so many hungred prisoners. Therefore impound not your Bees, so profitable and free a creature.

Taking of Bees. Let none stand aboue three yeares, else the combes will be blacke and knotty, your honey will be thinne and vncleanly: and if any cast after three yeares, it is such as haue swarmes, and old Bees kept all together, which is great losse. Smoaking with ragges, rozen, or brimstone, many vse: some vse drowning in a tub of cleane water, and the water well brewde, will be good botchet. Drawe out your spelkes immediatly with a paire of pinchars, lest the wood grow soft and swell, and so will not be drawne, then must you cut your Hiue.

Straining Honey. Let no fire come neere your hony, for fire softeneth the waxe and drosse, and makes them runne with the hony. Fire softneth, weakeneth, and hindereth hony for purging. Breake your combes small (when the dead empty combes are parted from the loaden combes) into a siue, borne ouer a great bowle, or vessell, with two staues, and so let it runne two or three dayes. The sooner you tunne it vp, the better will it purge. Runne your swarme honey by it selfe, and that shall be your best. The elder your hiues are, the worse is your honey.

Vessels. Vsuall vessels are of clay, but after wood be satiated with honey (for it will leake at first: for honey is maruellously searching, the thicke, and therefore vertuous.) I vse it rather because it will not breake so soone, with fals, frosts, or otherwise, and greater vessels of clay will hardly last.

When you vse your honey, with a spoone take off the skin which it hath put vp.

And it is worth the regard, that bees thus vsed, if you haue but forty stockes, shall yeeld you more commodity cleerely than forty acres of ground. And thus much may suffice, to make good House-wiues loue and haue good Gardens and Bees.

Deo Laus.

FINIS.

The Contents of the Countrey *House-wifes Garden.*

Chap. 1.	The Soyle.	Pag. 77	Bee-house.	p. 98.
Chap. 2.	Site.	p. 78	Hiues.	p. 100.
Chap. 3.	Forme.	p. 79	Hiuing of Bees.	p. ibid.
Chap. 4.	Quantity.	p. 85	Spelkes.	p. 101.
Chap. 5.	Fences.	p. ibid.	Catching.	p. 102.
Chap. 6.	Two Gardens.	p. 86	Clustering.	p. 103.
Chap. 7.	Diuision of herbs.	p. 88	Droanes.	p. 104.
Chap. 8.	The Husbandry of herbes.	p. ibid.	Annoyances.	p. 105.
Chap. 9.	Generall rules.	p. 96	Taking of Bees.	p. 106.
Chap. 10.	The Husbandry of Bees.	p. 98	Straining honey.	p. ibid.
			Vessels.	p. ibid.

A MOST PROFITABLE NEWE TREATISE,
From approued experience of the art *of propagating Plants: by* Simon Harward.

Chap. 1.
The Art of propagating Plants.

1. There are foure sorts of Planting, or propagating, as in laying of shootes or little branches, whiles they are yet tender in some pit made at their foote, as shall be sayd hereafter, or vpon a little ladder or Basket of earth, tyed to the bottome of the branch, or in boaring a Willow thorow, and putting the branch of the tree into the hole, as shall be fully declared in the Chapter of Grafting.

2. There are likewise seasons to propagate in; but the best is in the Spring, and *March*, when the trees are in the Flower, and doe begin to grow lusty. The young planted Siens or little Grafts must be propagated in the beginning of Winter, a foot deepe in the earth, and good manure mingled amongst the earth, which you shall cast forth of the pit, wherein you meane to propagate it, to tumble it in vpon it againe. In like manner your superfluous Siens, or little Plants must be cut close by the earth, when as they grow about some small Impe, which we meane to propagate, for they would doe nothing but rot. For to propagate, you must digge the earth round about the tree, that so your rootes may be laid in a manner halfe bare. Afterward draw into length the pit on that side where you meane to propagate, and according as you perceiue that the roots will be best able to yeeld, and be gouerned in the same pit, to vie them, and that with all gentlenesse, and stop close your Siens, in such sort, as that the wreath which is in the place where it is grafted, may be a little lower then the Siens of the new Wood, growing out of the earth, euen so high as it possible may be. If the trees that you would propagate be somewhat thicke, and thereby the harder to ply, and somewhat stiffe to lay in the pit: then you may wet the stocke almost to the midst, betwixt the roote and the wreathing place, and so with gentle handling of it, bow downe into the pit the

wood which the grafts haue put forth, and that in as round a compasse as you can, keeping you from breaking of it: afterward lay ouer the cut, with gummed Waxe, or with grauell and sand.

Chap. 2.
Grafting in the Barke.

Grafting in the Barke, is vsed from mid-*August*, to the beginning of Winter, and also when the Westerne winde beginneth to blow, being from the 7. of *February*, vnto 11. of *Iune*. But there must care be had, not to graffe in the barke in any rainy season, because it would wash away the matter of ioyning the one and the other together, and so hinder it.

3. Grafting in the budde, is vsed in the Summer time, from the end of *May*, vntill *August*, as being the time when the trees are strong and lusty, and full of sap and leaues. To wit, in a hot Countrey, from the midst of *Iune*, vnto the midst of *Iuly*: but cold Countries, to the midst of *August*, after some small showres of Raine.

If the Summer be so exceeding dry, as that some trees doe withhold their sap, you must waite the time till it doe returne.

Graft from the full of the Moone, vntill the end of the old.

You may graft in a Cleft, without hauing regard to the Raine, for the sap will keepe it off.

You may graft from mid-*August*, to the beginning of *Nouember*: Cowes dung with straw doth mightily preserue the graft.

It is better to graft in the euening, then the morning.

The furniture and tooles of a Grafter, are a Basket to lay his Grafts in, Clay, Grauell, Sand, or strong Earth, to draw ouer the plants clouen: Mosse, Woollen clothes, barkes of Willow to ioyne to the late things and earth before spoken, and to keepe them fast: Oziers to tye againe vpon the barke, to keepe them firme and fast: gummed Wax, to dresse and couer the ends and tops of the grafts newly cut, that so the raine and cold may not hurt them, neither yet the sap rising from belowe, be constrained to returne againe vnto the shootes. A little Sawe or hand Sawe, to sawe off the stocke of the plants, a little Knife or Pen-knife to graffe, and to cut and sharpen the grafts, that so the barke may not pill nor be broken; which often commeth to passe when the graft is full of sap. You shall cut the graffe so long, as that it may fill the cliffe of the plant, and therewithall it must be left thicker on the barke-side, that so it may fill vp both the cliffe and other incisions, as any need is to be made, which must be alwaies well ground, well burnished without all rust. Two wedges, the one broad for thicke trees, the other narrow for lesse and tender trees, both of them of box, or some other hard and smooth wood, or steele, or of very hard iron, that so they may need lesse labour in making them sharpe.

A little hand-Bill to set the plants at more liberty, by cutting off superfluous boughs, helu'd of Iuory, Box, or Brazell.

Chap. 3.
Grafting in the cleft.

The manner of grafting in a cleft, to wit, the stocke being clou'd, is proper not onely to trees, which are as great as a mans legs or armes, but also to greater. It is true that in as much as the trees cannot easily be clouen in their stocke, that therefore it is expedient to make incision in some one of their branches, and not in the maine body, as we see to be practised in great Apple trees, and great Peare-trees, and as we haue already declared heretofore.

To graft in the cleft, you must make choise of a graft that is full of sap and iuyce, but it must not bee, but till from after *Ianuary* vntill *March*: And you must not thus graft in any tree that is already budded, because a great part of the iuyce and sap would be already mounted vp on high, and risen to the top, and there dispersed and scattered hither and thither, into euery sprigge and twigge, and vse nothing welcome to the graft.

You must likewise be resolued not to gather your graft the day you graft in, but ten or twelue dayes before: for otherwise, if you graft it new gathered, it will not be able easily to incorporate itselfe with the body and stocke, where it shall be grafted; because that some part of it will dry, and by this meanes will be a hinderance in the stocke to the rising vp of the sap, which it should communerate vnto the graft, for the making of it to put forth, and whereas this dried part will fall a crumbling, and breaking thorow his rottennesse, it will cause to remaine a concauity, or hollow place in the stock, which will be an occasion of a like inconuenience to befall the graft. Moreouer, the graft being new and tender, might easily be hurt of the bands, which are of necessity to be tyed about the Stocke, to keepe the graft firme and fast. And you must further see, that your Plant was not of late remoued, but that it haue already fully taken root.

When you are minded to graft many grafts into one cleft, you must see that they be cut in the end all alike.

7. See that the grafts be of one length, or not much differing, and it is enough, that they haue three or foure eylets without the wrench when the Plant is once sawed, and lopped of all his small Siens and shootes round about, as also implyed of all his branches, if it haue many: then you must leaue but two at the most, before you come to the cleauing of it: then put to your little Saw, or your knife, or other edged toole that is very sharpe, cleaue it quite thorow the middest, in gentle and soft sort: First, tying the Stocke very sure, that so it may not cleaue further then is need: and then put to your Wedges into the cleft vntill such time as you haue set in your grafts, and in cleauing of it, hold the knife with the one hand, and the tree with the other, to helpe to keepe it from cleauing too farre. Afterwards put in your wedge of Boxe or Brazill, or bone at the small end, that so you may the better take it out againe, when you haue set in your grafts.

8. If the Stocke be clouen, or the Barke loosed too much from the wood: then cleaue it downe lower, and set your grafts in, and looke that their incision bee fit, and very iustly answerable to the cleft, and that the two saps, first, of the Plant and graft, be right and euen set one against the other, and so handsomely fitted, as that there may not be the least appearance of any cut or cleft. For if they doe not thus lumpe one with another, they will neuer take one with another, because they cannot worke their seeming matter, and as it were cartilaguous glue in conuenient sort or manner, to the gluing of their ioynts together. You must likewise beware, not to make your cleft ouerthwart the pitch, but somewhat aside.

The barke of your Plant being thicker then that of your Graft, you must set the graft so much the more outwardly in the cleft, that so the two saps may in any case be ioyned, and set right the one with the other but the rinde of the Plant must be somewhat more out, then that of the grafts on the clouen side.

9. To the end that you may not faile of this worke of imping, you must principally take heed, not to ouer-cleaue the Stockes of your Trees. But before you widen the cleft of your wedges, binde, and goe about the Stocke with two or three turnes, and that with an Ozier, close drawne together, vnderneath the same place, where you would haue your cleft to end, that so your Stocke cleaue not too farre, which is a very vsuall cause of the miscarrying of grafts, in asmuch as hereby the cleft standeth so wide and open, as that it cannot be shut, and so not grow together againe; but in the meane time spendeth it selfe, and breatheth out all his life in that place, which is the cause that the Stocke and the Graft are both spilt. And this falleth out most often in Plum-trees, & branches of trees. You must be careful so to ioyne the rinds of your grafts, and Plants, that nothing may continue open, to the end that the wind, moisture of the clay or raine, running vpon the grafted place, do not get in: when the plant cloueth very straight, there is not any danger nor hardnesse in sloping downe the Graft. 10. If you leaue it somewhat vneuen, or rough in some places, so that the saps both of the one and of the other may the better grow, and be glued together, when your grafts are once well ioyned to your Plants, draw out your wedges very softly, lest you displace them againe, you may leaue there within the cleft some small end of a wedge of greene wood, cutting it very close with the head of the Stocke: Some cast glue into the cleft, some Sugar, and some gummed Waxe.

If the Stocke of the Plant whereupon you intend to graft, be not so thicke as your graft, you shall graft it after the fashion of a Goates foot, 11. make a cleft in the Stocke of the Plant, not direct, but byas, & that smooth and euen, not rough: then apply and make fast there-to, the graft withall his Barke on, and answering to the barke of the Plant. This being done, couer the place with the fat earth and mosse of the Woods tyed together with a strong band: sticke a pole of Wood by it, to keepe it stedfast.

Chap. 4.
Grafting like a Scutcheon.

In grafting after the manner of a Scutcheon, you shall not vary nor differ much from that of the Flute or Pipe, saue only that the Scutcheon-like graft, hauing one eyelet, as the other hath yet the wood of the tree whereupon the Scutcheon-like graft is grafted, hath not any knob, or budde, as the wood whereupon the graft is grafted, after the manner of a pipe.

12. In Summer when the trees are well replenished with sap, and that their new Siens begin to grow somewhat hard, you shall take a shoote at the end of the chiefe branches of some noble and reclaimed tree, whereof you would faine haue some fruit, and not many of his old store or wood, and from thence raise a good eylet, the tayle and all thereof to make your graft. But when you choose, take the thickest, and grossest, diuide the tayle in the middest, before you doe any thing else, casting away the leafe (if it be not a Peare plum-tree: for that would haue two or three leaues) without remouing any more of the said tayle: afterward with the point of a sharpe knife, cut off the Barke of the said shoote, the patterne of a shield, of the length of a nayle.

13. In which there is onely one eylet higher then the middest together, with the residue of the tayle which you left behinde: and for the lifting vp of the said graft in Scutcheon, after that you haue cut the barke of the shoote round about, without cutting of the wood within, you must take it gently with your thumbe, and in putting it away you must presse vpon the wood from which you pull it, that so you may bring the bud and all away together with the Scutcheon: for if you leaue it behinde with the wood, then were the Scutcheon nothing worth. You shall finde out if the Scutcheon be nothing worth, if looking within when it is pulled away from the wood of the same sute, you finde it to haue a hole within, but more manifestly, if the bud doe stay behind in the VVood, which ought to haue beene in the Scutcheon.

Thus your Scutcheon being well raised and taken off, hold it a little by the tayle betwixt your lips, 14. without wetting of it, euen vntill you haue cut the Barke of the tree where you would graft it, and looke that it be cut without any wounding of the wood within, after the manner of a crutch, but somewhat longer then the Scutcheon that you haue to set in it, and in no place cutting the wood within; after you haue made incision, you must open it, and make it gape wide on both sides, but in all manner of gentle handling, and that with little Sizers of bone, and separating the wood and the barke a little within, euen so much as your Scutcheon is in length and breadth: you must take heed that in doing hereof, you do not hurt the bark.

This done take your Scutcheon by the end, and your tayle which you haue left remaining, and put into your incision made in your tree, 15. lifting vp softly your two sides of the incision with your said Sizers of bone, and cause the said Scutcheon to ioyne, and lye as close as may be, with the wood of the tree, being cut, as aforesaid, in waying a little vpon the end of your rinde: so cut and let the vpper part of your Scutcheon lye close vnto the vpper end of your incision, or barke of your said tree: afterward binde your Scutcheon about with a band of Hempe, as thicke as a pen or a quill, more or lesse, according as your tree is small or great, taking the same Hempe in the middest, to the end that either part of it may performe a like seruice; and wreathing and binding of the said Scutcheon into the incision of a tree, and it must not be tyed too strait, for that would keepe it from taking the ioyning of the one sap to the other, being hindred thereby, and neither the Scutcheon, nor yet the Hempe must be moist or wet: and the more iustly to binde them together, begin at the backe side of the Tree, right ouer against the middest of the incision, and from thence come forward to ioyne them before, aboue the eylet and tayle of the Scutcheon, crossing your band of Hempe, so oft as the two ends meet, and from thence returning backe againe, come about and tye it likewise vnderneath the eylets: and thus cast about your band still backward and forward, vntill the whole cleft of the incision be couered aboue and below with the said Hempe, the eylet onely excepted, and his tayle which must not be couered at all; 17. his tayle will fall away one part after another, and that shortly after the ingrafting, if so be the Scutcheon will take. Leaue your trees and Scutcheons thus bound, for the space of one moneth, and the thicker, a great deale longer time. Afterward looke them ouer, and if you perceiue them to grow together, vntye them, or at the leastwise cut the Hempe behinde them, and leaue them vncouered. Cut also your branch two or three fingers aboue that, so the impe may prosper the better: and thus let them remaine till after Winter, about the moneth of *March*, and *Aprill*.

If you perceiue that your budde of your Scutcheon doe swell and come forward: then cut off the tree three fingers or thereabouts, aboue the Scutcheon: 18. for if it be cut off too neere the Scutcheon, at such time as it putteth forth his first blossome, it would be a meanes greatly to hinder the flowring of it, and cause also that it should not thriue and prosper so well after that one yeere is past, and that the shoote beginneth to be strong: beginning to put forth the second bud and blossome, you must goe forward to cut off in byas-wise the three fingers in the top of the tree, which you left there, when you cut it in the yeere going before, as hath beene said.

19. When your shoote shall haue put foorth a great deale of length, you must sticke down there, euen hard ioyned thereunto, little stakes, tying them together very gently and easily; and these shall stay your shootes and prop them vp, letting the winde from doing any harme vnto them. Thus you may graft white Roses in red, and red in white. Thus you may graft two or three Scutcheons: prouided that they be all of one side: for they will not be set equally together in height because then they would bee all staruelings, neither would they be directly one ouer another; 20. for the lower would stay the rising vp of the sap of the tree, and so those aboue should consume in penury, and vndergoe the aforesaid inconuenience. You must note, that the Scutcheon which is gathered from the Sien of a tree whose fruite is sowre, must be cut in square forme, and not in the plaine fashion of a Scutcheon. It is ordinary to graffe the sweet Quince tree, bastard Peach-tree, Apricock-tree, Iuiube-tree, sowre Cherry tree, sweet Cherry-tree, and Chestnut tree, after this fashion, howbeit they might be grafted in the cleft more easily, and more profitably; although diuers be of contrary opinion, as thus best: Take the grafts of sweet Quince tree, and bastard Peach-tree, or the fairest wood, and best fed that you can finde, growing vpon the wood of two yeeres old, 21. because the wood is not so firme nor solid as the others, and you shall graffe them vpon small Plum-tree stocks, being of the thicknes of ones thumbe; these you shall cut after the fashion of a Goats foot: you shall not goe about to make

the cleft of any more sides then one, being about a foot high from the ground; you must open it with your small wedge: and being thus grafted, it will seeme to you that it is open but of one side; afterward you shall wrap it vp with a little Mosse, putting thereto some gummed Wax, or clay, and binde it vp with Oziers to keepe it surer, because the stocke is not strong enough it selfe to hold it, and you shall furnish it euery manner of way as others are dealt withall: this is most profitable.

The time of grafting.

All moneths are good to graft in, (the moneth of *October* and *Nouember* onely excepted). But commonly, graft at that time of the Winter, when sap beginneth to arise.

In a cold Countrey graft later, and in a warme Countrey earlier.

The best time generall is from the first of *February*, vntill the first of *May*.

The grafts must alwaies be gathered, in the old of the Moone.

For grafts choose shootes of a yeere old, or at the furthermost two yeeres old.

If you must carry grafts farre, pricke them into a Turnep newly gathered, or lay earth about the ends.

If you set stones of Plummes, Almonds, Nuts, or Peaches: First let them lye a little in the Sunne, and then steepe them in Milke or Water, three or foure dayes before you put them into the earth.

Dry the kernels of Pippins, and sow them in the end of *Nouember*.

The stone of a Plum-tree must be set a foot deepe in *Nouember*, or *February*.

The Date-stone must be set the great end downwards, two cubits deepe in the earth, in a place enriched with dung.

The Peach-stone would be set presently after the Fruit is eaten, some quantity of the flesh of the Peach remaining about the stone.

If you will haue it to be excellent, graft it afterward vpon an Almond tree.

The little Siens of Cherry-trees, grown thicke with haire, rots, and those also which doe grow vp from the rootes of the great Cherry-trees, being remoued, doe grow better and sooner then they which come of stones: but they must be remoued and planted while they are but two or three yeeres old, the branches must be lopped.

The Contents of the Art of *Propagating Plants.*

The Art of propagating Plants.	page 109.	*Inoculation in the Barke.*
Grafting in the Barke.	p. 111.	*Emplaister-wise grafting.*
Grafting in the cleft.	p. 113.	*To pricke stickes to beare the first yeere.*
Grafters Tooles.		*To haue Cherries or Plums without stones.*
Time of planting & seting.		*To make Quinces great.*
Time of grafting.		*To set stones of Plummes.*
How to cut the stumps in grafting.		*Dates, Nut, and Peaches.*
Sprouts and imps: how gathered.		*To make fruit smell well.*
Grafting like a Scutcheon.	p. 116.	*To plant Cherry-trees.*

THE HVSBAND MANS FRVITEFVLL ORCHARD.

> For the true ordering of all sorts of
> *Fruits in their due seasons; and how double*
> increase commeth by care in gathering
> *yeere after yeare: as also the best way*
> of carriage by land or by water:
> *With their preseruation for*
> longest continuance.

Cherries. Of all stone Fruit, Cherries are the first to be gathered: of which, though we reckon foure sorts; *English, Flemish, Gascoyne* and *Blacke*, yet are they reduced to two, the early, and the ordinary: the earely are those whose grafts came first from *France* and *Flanders*, and are now ripe with vs in *May*: the ordinary is our owne naturall Cherry, and is not ripe before *Iune*; they must be carefully kept from Birds, either with   nets, noise, or other industry.

Gathering of Cheries. They are not all ripe at once, nor may be gathered at once, therefore with a light Ladder, made to stand of it selfe, without hurting the boughes, mount to the tree, and with a gathering hooke, gather those which be full ripe, and put them into your Cherry-pot, or Kybzey hanging by your side, or vpon any bough you please, and be sure to breake no stalke, but that the cherry hangs by; and pull them gently, lay them downe tenderly, and handle them as little as you can.

To carry Cherries. For the conueyance or portage of Cherries, they are best to be carried in broad Baskets like siues, with smooth yeelding bottomes, onely two broad laths going along the bottome: and if you doe transport them by ship, or boate, let not the siues be fil'd to the top, lest setting one vpon another, you bruise and hurt the Cherries: if you carry by horse-backe, then panniers well lined with Fearne, and packt full and close is the best and safest way.

Other stone-fruit. Now for the gathering of all other stone-fruite, as Nectarines, Apricockes, Peaches, Peare-plumbes, Damsons, Bullas, and such like, although in their seuerall kinds, they seeme not to be ripe at once on one tree: yet when any is ready to drop from the tree, though the other seeme hard, yet they may also be gathered, for they haue receiued the full substance the tree can giue them; and therefore the day being faire, and the dew drawne away; set vp your Ladder, and as you gathered your Cherries, so gather them: onely in the bottomes of your large siues, where you part them, you shall lay Nettles, and likewise in the top, for that will ripen those that are most vnready.

Gathering of Peares. In gathering of Peares are three things obserued;   to gather for expence, for transportation, or to sell to the Apothecary. If for expence, and your owne vse, then gather them as soone as they change, and are as it were halfe ripe, and no more but those which are changed, letting the rest hang till they change also: for thus they will ripen kindely, and not rot so soone, as if they were full ripe at the gathering. But if your Peares be to be transported farre either by Land or Water, then pull one from the tree, and cut it in the middest, and if you finde it hollow about the choare, and the kernell a large space to lye in: although no Peare be ready to drop from the tree, yet then they may be gathered, and then laying them on a heape one vpon another, as of necessity they must be for transportation, they will ripen of themselues, and eate kindly: but gathered before, they will wither, shrinke and eate rough, losing not onely their taste, but beauty.

Now for the manner of gathering; albeit some climb into the trees by the boughes, and some by Ladder, yet both is amisse: the best way is with the Ladder before spoken of, which standeth of it selfe, with a basket and a line, which being full, you must gently let downe, and keeping the string still in your hand, being emptied, draw it vp againe, and so finish your labour, without troubling your selfe, or hurting the tree.

Gathering of Apples. Now touching the gathering of Apples, it is to be done according to the ripening of the fruite; your Summer apples first, and the Winter after.

For Summer fruit, when it is ripe, some will drop from the tree, and birds will be picking at them: But if you cut one of the greenest, and finde it as was   shew'd you before of the Peare: then you may gather them, and in the house they will come to their ripenesse and perfection. For your Winter fruit, you shall know the ripenesse by the obseruation before shewed; but it must be gathered in a faire, Sunny, and dry day, in the waine of the Moone, and no Wind in the East, also after the deaw is gone away: for the least wet or moysture will make them subiect to rot and mildew: also you must haue an apron to gather in, and to empty into the great baskets, and a hooke to draw the boughes vnto you, which you cannot reach with your hands at ease: the apron is to be an Ell euery way, loopt vp to your girdle, so as it may serue for either hand without any trouble: and when it is full, vnloose one of your loopes, and empty it gently into the great basket, for in throwing them downe roughly, their owne stalkes may pricke them; and those which are prickt, will euer rot. Againe, you must gather your fruit cleane without leaues or brunts, because the one hurts the tree, for euery brunt would be a stalke for fruit to grow vpon: the other hurts the fruit by bruising, and pricking it as it is layd together, and there is nothing sooner rotteth fruit, then the greene and withered leaues lying amongst them; neither must you gather them without any stalke at all: for such fruit will begin to rot where the stalke stood.

To vse the fallings. For such fruit as falleth from the trees, and are not gathered, they must not be layd with the gathered fruit: and of fallings there are two sorts, one that fals through ripenesse, and they are best, and may be kept to bake or roast; the other windfals, and before they are ripe, and they must be spent as they are gathered, or else they will wither and come to nothing: and therefore it is not good by any meanes to beate downe fruit with Poales, or to carrie them in Carts loose and iogging or in sacks where they may be bruised.

Carriage of fruit. When your fruit is gathered, you shall lay them in deepe Baskets of Wicker, which shall containe foure or sixe bushels, and so betweene two men, carry them to your Apple-Loft, and in shooting or laying them downe, be very carefull that it be done with all gentlenesse, and leasure, laying euery sort of fruit seuerall by it selfe: but if there be want of roome hauing so many sorts that you cannot lay them seuerally, then such some fruite as is neerest in taste and colour, and of Winter fruit, such as will taste alike, may if need require, be laid together, and in time you may separate them, as shall bee shewed hereafter. But if your fruit be gathered faire from your Apple-Loft, them must the bottomes of your Baskets be lined with greene Ferne, and draw the stuborne ends of the same through the Basket, that none but the soft leafe may touch the fruit, and likewise couer the tops of the Baskets with Ferne also, and draw small cord ouer it, that the Ferne may not fall away, nor the fruit scatter out, or iogge vp and downe: and thus you may carry fruit by Land or by Water, by Boat, or Cart, as farre as you please: and the Ferne doth not onely keepe them from bruising, but also ripens them, especially Peares. When your fruit is brought to your Apple-Loft or store house, if you finde them not ripened enough, then lay them in thicker heapes vpon Fearne, and couer them with Ferne also: and when they are neere ripe, then vncouer them, and make the heapes thinner, so as the ayre may passe thorow them: and if you will not hasten the ripening of them, then lay them on the boords without any Fearne at all. Now for Winter, or long lasting Peares, they may be packt either in Ferne or Straw, and carried whither you please; and being come to the iourneys end must be laid vpon sweet straw; but beware the roome be not too warme, nor windie, and too cold, for both are hurtfull: but in a temperate place, where they may haue ayre, but not too much.

Of Wardens. Wardens are to be gathered, carried, packt, and laid as Winter Peares are.

Of Medlers. Medlers are to be gathered about *Michaelmas*, after a frost hath toucht them; at which time they are in their full growth, and will then be dropping from the tree, but neuer ripe vpon the tree. When they are gathered, they must be laid in a basket, siue, barrell, or any such caske, and wrapt about with woollen cloths, vnder, ouer, and on all sides, and also some waight laid vpon them, with a boord betweene: for except they be brought into a heat, they will neuer ripen kindly or taste well.

Now when they haue laine till you thinke some of them be ripe, the ripest, still as they ripen, must be taken from the rest: therefore powre them out into another siue or basket leasurely, that so you may well finde them that be ripest, letting the hard one fall into the other basket, and those which be ripe laid aside: the other that be halfe ripe, seuer also into a third siue or basket: for if the ripe and halfe ripe be kept together, the one will be mouldy, before the other be ripe: And thus doe, till all be throughly ripe.

Of Quinces. Quinces should not be laid with other fruite; for the sent is offensiue both to other fruite, and to those that keepe the fruite or come amongst them: therefore   lay them by themselues vpon sweet strawe, where they may haue ayre enough: they must be packt like Medlers, and gathered with Medlers.

To packe Apples. Apples must be packt in Wheat or Rye-straw, and in maunds or baskets lyned with the same, and being gently handled, will ripen with such packing and lying together. If seuerall sorts of apples be packt in one maund or basket, then betweene euery sort, lay sweet strawe of a pretty thicknesse.

Emptying and laying apples. Apples must not be powred out, but with care and leasure: first, the straw pickt cleane from them, and then gently take out euery seuerall sort, and place them by themselues: but if for want of roome you mixe the sorts together, then lay those together that are of equall lasting; but if they haue all one taste, then they need no separation. Apples that are not of the like colours should not be laid together, and if any such be mingled, let it be amended, and those which are first ripe, let them be first spent; and to that end, lay those apples together, that are of one time ripening: and thus you must vse Pippins also, yet will they endure bruises better then other fruit, and whilst they are greene will heale one another.

Difference in Fruit. Pippins though they grow of one tree, and in one ground, yet some will last better then other some, and some will bee bigger then others of the same kinde, according as they haue more or lesse of the Sunne, or more or lesse of the droppings of the trees or vpper branches: therefore let euery one make most of that fruite which is fairest, and longest lasting. Againe, the largenesse and goodnesse of fruite consists in the age of the tree: for as the tree increaseth, so the fruite increaseth in bignesse, beauty, taste, and firmnesse: and otherwise, as it decreaseth.

Transporting fruit by water. If you be to transport your fruit farre by water, then prouide some dry hogges-heads or barrells, and packe in your apples, one by one with your hand, that no empty place may be left, to occasion sogging; and you must line your vessell at both ends with fine sweet straw; but not the sides, to auoid heat: and you must bore a dozen holes at either end, to receiue ayre so much the better; and by no meanes let them take wet. Some vse, that transport beyond seas, to shut the fruite vnder hatches vpon straw: but it is not so good, if caske may be gotten.

When not to transport fruit. It is not good to transport fruite in *March*, when the wind blowes bitterly, nor in frosty weather, neither in the extreme heate of Summer.

To conuay small store of fruit. If the quantity be small you would carry, then you may carry them in Dossers or Panniers, prouided they be euer filled close, and that Cherries and Peares be lined with greene Fearne, and Apples with sweete straw; and that, but at the bottomes and tops, not on the sides.

Roomes for fruite. Winter fruite must lye neither too hot, nor too cold; too close, nor too open: for all are offensiue. A lowe roome or Cellar that is sweet, and either boorded or paued, and not too close, is good, from *Christmas* till *March*: and roomes that are seeled ouer head, and from the ground, are good from *March* till *May*: then the Cellar againe, from *May* till *Michaelmas*. The apple loft would be seeled or boorded, which if it want, take the longest Rye-straw, and raise it against the walles, to make a fence as high as the fruite lyeth; and let it be no thicker then to keepe the fruite from the wall, which being moyst, may doe hurt, or if not moist, then the dust is offensiue.

Sorting of Fruit. There are some fruite which will last but vntill *Allhallontide*: they must be laid by themselues; then those which will last till *Christmas*, by themselues: then those which will last till it be *Candlemas*, by themselues: those that will last till *Shrouetide*, by themselues: and Pippins, Apple-Iohns, Peare-maines, and Winter-Russettings, which will last all the yeere by themselues.

Now if you spy any rotten fruite in your heapes, pick them out, and with a Trey for the purpose, see you turne the heapes ouer, and leaue not a tainted Apple in them, diuiding the hardest by themselues, and the broken skinned by themselues to be first spent, and the rotten ones to be cast away; and euer as you turne them, and picke them, vnder-lay them with fresh straw: thus shall you keepe them safe for your vse, which otherwise would rot suddenly.

Times of stirring fruit. Pippins, Iohn Apples, Peare maines, and such like long lasting fruit, need not to be turned till the weeke before *Christmas*, vnlesse they be mixt with other of a riper kind, or that the fallings be also with them, or much of the first straw left amongst them: the next time of turning is at *Shroue-tide*, and after that, once a moneth till *Whitson-tide*; and after that, once a fortnight; and euer in the turning, lay your heapes lower and lower, and your straw very thinne: prouided you doe none of this labour in any great frost, except it be in a close Celler. At euery thawe, all fruit is moyst, and then they must not be touched: neither in rainy weather, for then they will be danke also: and therefore at such seasons it is good to set open your windowes, and doores, that the ayre may haue free passage to dry them, as at nine of the clocke in the fore-noone in Winter; and at six in the fore-noone, and at eight at night in Summer: onely in *March*, open not your windowes at all.

All lasting fruite, after the middest of *May*, beginne to wither, because then they waxe dry, and the moisture gone, which made them looke plumpe: they must needes wither, and be smaller; and nature decaying, they must needes rot. And thus much touching the ordering of fruites.

FINIS.

LONDON,
Printed by *Nicholas Okes* for Iohn Harison, at the
golden Vnicorne in Pater-noster-row. 1631.

www.ingramcontent.com/pod-product-compliance
Lightning Source LLC
Chambersburg PA
CBHW031431210526
45464CB00005B/2153